Lecture Notes in Physics

Springer

Berlin
Heidelberg
New York
Barcelona
Hong Kong
London
Milan
Paris
Tokyo

Physics and Astronomy

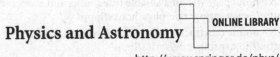

http://www.springer.de/phys/

Editorial Policy

The series *Lecture Notes in Physics* (LNP), founded in 1969, reports new developments in physics research and teaching -- quickly, informally but with a high quality. Manuscripts to be considered for publication are topical volumes consisting of a limited number of contributions, carefully edited and closely related to each other. Each contribution should contain at least partly original and previously unpublished material, be written in a clear, pedagogical style and aimed at a broader readership, especially graduate students and nonspecialist researchers wishing to familiarize themselves with the topic concerned. For this reason, traditional proceedings cannot be considered for this series though volumes to appear in this series are often based on material presented at conferences, workshops and schools (in exceptional cases the original papers and/or those not included in the printed book may be added on an accompanying CD ROM, together with the abstracts of posters and other material suitable for publication, e.g. large tables, colour pictures, program codes, etc.).

Acceptance

A project can only be accepted tentatively for publication, by both the editorial board and the publisher, following thorough examination of the material submitted. The book proposal sent to the publisher should consist at least of a preliminary table of contents outlining the structure of the book together with abstracts of all contributions to be included.

Final acceptance is issued by the series editor in charge, in consultation with the publisher, only after receiving the complete manuscript. Final acceptance, possibly requiring minor corrections, usually follows the tentative acceptance unless the final manuscript differs significantly from expectations (project outline). In particular, the series editors are entitled to reject individual contributions if they do not meet the high quality standards of this series. The final manuscript must be camera-ready, and should include both an informative introduction and a sufficiently detailed subject index.

Contractual Aspects

Publication in LNP is free of charge. There is no formal contract, no royalties are paid, and no bulk orders are required, although special discounts are offered in this case. The volume editors receive jointly 30 free copies for their personal use and are entitled, as are the contributing authors, to purchase Springer books at a reduced rate. The publisher secures the copyright for each volume. As a rule, no reprints of individual contributions can be supplied.

Manuscript Submission

The manuscript in its final and approved version must be submitted in camera-ready form. The corresponding electronic source files are also required for the production process, in particular the online version. Technical assistance in compiling the final manuscript can be provided by the publisher's production editor(s), especially with regard to the publisher's own Latex macro package which has been specially designed for this series.

Online Version/ LNP Homepage

LNP homepage (list of available titles, aims and scope, editorial contacts etc.):
http://www.springer.de/phys/books/lnpp/

LNP online (abstracts, full-texts, subscriptions etc.):
http://link.springer.de/series/lnpp/

Daniel Benest   Claude Froeschlé   (Eds.)

Singularities in Gravitational Systems

Applications to Chaotic Transport in the Solar System

Springer

Editor

Daniel Benest
Claude Froeschlé
O.C.A. Observatoire de Nice
B.P. 4229
06304 Nice Cedex 4, France

Cover picture: see figure 5, page 97, contribution by J. Waldvogel in this volume

Library of Congress Cataloging-in-Publication Data.

Die Deutsche Bibliothek - CIP-Einheitsaufnahme

Singularities in gravitational systems : applications to chaotic transport
in the solar system / Daniel Benest : Claude Froeschle (ed.). - Berlin ;
Heidelberg ; New York ; Barcelona ; Hong Kong ; London ; Milan ; Paris ;
Tokyo : Springer, 2002
    (Lecture notes in physics ; Vol. 590)
    (Physics and astronomy online library)

ISSN 0075-8450
ISBN 978-3-642-07844-6
e-ISBN 978-3-540-48009-9

Springer-Verlag Berlin Heidelberg New York
a member of BertelsmannSpringer Science+Business Media GmbH

http://www.springer.de

© Springer-Verlag Berlin Heidelberg 2002
Softcover reprint of the hardcover 1st edition 2002

Cover design: *design & production*, Heidelberg

Preface

The theory of chaos invades a large part of modern physics (celestial mechanics, fluid mechanics, particle accelerators, solid mechanics, etc.), together with other branches of knowledge (biology, ecology, economics, etc.). In particular, important results have been obtained in astronomy, especially in the study of gravitational systems.

In this domain, one kind of chaos is due to interactions between several resonances, which are at the origin of the weak chaos detected in our Solar System, particularly for the inner planets and, for example, in the attitude variations of Mars; let us recall that our Earth, thanks to the influence of the Moon, does not suffer attitude variations (which would be catastrophic for the stability of the climate) as strong as the latter planet. Theoretically speaking, the KAM theorem establishes the persistence of invariant tori in weakly perturbed Hamiltonian systems; besides, the Nekhorochev theorem allows the confinement over very long times, under some constraints, of weakly chaotic orbits.

But another kind of chaos, completely independent from interactions between resonances, is due to close encounters between celestial bodies, and is responsible for, for example, rapid transfer of "killing" asteroids which cross the orbits of telluric planets and can hurt them (let us recall, for example, the Cretaceous-Tertiary event); this can be the cause of ejection of comets and some asteroids away from the Solar System, too. Moreover, we should remind ourselves of the use of these close encounters by space missions, during which energy is given to spacecrafts through "rebounds" on planets (e.g. for the Cassini mission). In the equations of motion, denominators equal to the square of mutual distances between the bodies become, during such close encounters, very small and induce *singularities*; one of the solutions found for "rubbing out" these singularities during the integration is called *regularization*, which uses transformations on space and time. More recently, Öpik's works have allowed modeling of this close-encounter-induced chaos, and have been applied to the study of meteor streams and to chaotic diffusion of particles in planetary rings. These were the topics lectured on during the Arc 2000 School in 2000 (organized together with our colleague Patrick Michel, also from Nice), and which consequently constitute the main focus of this book.

The early chapters introduce the mathematical methods used in the theory of singularities in gravitational systems (e.g. regularization).

The second part of the book develops the modelization techniques, in particular the elaboration of "mappings" in which the basic ingredient consists of introducing delta functions to represent close encounters as shocks.

Finally, the concluding chapters present the state of the art about the study of the diffusion of comets, wandering asteroids, meteors and planetary ring particles. Note that such studies are particularly relevant today, as the advances in modern observational instrumentation (LINEAR, Spacewatch, etc.) have lead to an enormous increase in the frequency of discovery of minor bodies in the Solar System.

General References

[1] Benest, D., Froeschlé, C. (eds.) (1990): *Les méthodes modernes de la Mécanique Céleste [Modern Methods of Celestial Mechanics]* (Goutelas 1989), Editions Frontières (C36).

[2] Benest, D., Froeschlé, C. (eds.) (1992): *Interrelations Between Physics and Dynamics for Minor Bodies in the Solar System* (Goutelas 1991), Editions Frontières (C49).

[3] Benest, D., Froeschlé, C. (eds.) (1994): *An Introduction to Methods of Complex Analysis and Geometry for Classical Mechanics and Non-Linear Waves* (Chamonix 1993), Editions Frontières.

[4] Benest, D., Froeschlé, C. (eds.) (1995): *Chaos and Diffusion in Hamiltonian Systems* (Chamonix 1994), Editions Frontières.

[5] Benest, D., Froeschlé, C. (eds.) (1998): *Impacts on Earth* (Goutelas 1994), Springer, Lecture Notes in Physics, Vol. 505.

[6] Benest, D., Froeschlé, C. (eds.) (1998): *Analysis and Modelling of Discrete Dynamical Systems – with Applications to Dynamical Astronomy* (Aussois 1996), Gordon and Breach – Advances in Discrete Mathematics and Applications, Vol. 1.

[7] Benest, D., Froeschlé, C. (eds.) (1999): *At the frontiers of chaotic dynamics of gravitational systems* (Arc 2000 1998), Special Issue of *Celestial Mechanics & Dynamical Astronomy*, Vol. 72 (1–2).

Nice,
February 2002

Daniel Benest
Claude Froeschlé

Table of Contents

Singularities, Collisions and Regularization Theory
Alessandra Celletti 1
1   Introduction 1
2   A world of singularities 3
3   Past and future collisions in the Solar System 4
4   Regularization theory 9
5   Triple collisions and central configurations 17
6   Chaotic diffusion: the inclined billiard 18
7   Noncollision singularities 20
References 23

The Levi–Civita, KS and Radial–Inversion Regularizing Transformations
Alessandra Celletti 25
1   Introduction 25
2   The two- and three-body problem 26
    2.1   The two-body problem 26
    2.2   The planar, circular, restricted three-body problem 26
3   The Levi–Civita regularization 29
    3.1   The two-body problem 29
    3.2   The planar, circular, restricted three-body problem 32
4   The Kustaanheimo–Stiefel regularization 36
    4.1   The Kustaanheimo–Stiefel transformation 36
    4.2   Canonicity of the KS-transformation 39
5   The radial–inversion transformation 44
References 48

The Birkhoff and B_3 Regularizing Transformations
Maria Gabriella Della Penna 49
1   Introduction 49
2   The Birkhoff regularization 49
3   The B_3 regularization 57
4   Numerical integration 60
References 62

Perturbative Methods in Regularization Theory

Corrado Falcolini 63

1   Introduction 63
2   Regularization procedure 63
    2.1   The general case 64
    2.2   The fictitious-time case 64
    2.3   The generalized eccentric anomaly case 65
3   Analytic perturbative methods 67
    3.1   Hamilton–Jacobi for the fictitious-time case 67
    3.2   Hamilton–Jacobi for the generalized eccentric anomaly case 68
    3.3   Series expansion 69
    3.4   First-order perturbation 70
References 71

Collisions and Singularities in the n-body Problem

Corrado Falcolini 72

1   Introduction 72
2   Non-collision singularities in Newtonian systems 72
    2.1   The n-body problem 73
3   The solution 75
    3.1   The example of Mather and McGehee 75
    3.2   The first example of Gerver 75
    3.3   The example of Xia 76
    3.4   The second example of Gerver 77
4   Multiple and simultaneous binary collisions 78
5   Open questions 79
References 79

Triple Collision and Close Triple Encounters

Jörg Waldvogel 81

1   Basics 81
    1.1   The general three-body problem: equations of motion,
          integrals of motion 82
    1.2   Angular momentum, Sundman's theory 83
2   Triple collision 86
    2.1   Homographic and homothetic solutions 86
    2.2   Central configurations 87
    2.3   Triple collision, Siegel's series................................. 90
3   The close triple encounter 94
    3.1   Singular perturbations... 95
    3.2   The triple-collision manifold.................................... 96
References 98

Dynamical and Kinetic Aspects of Collisions
Yves Elskens 101
1   Introduction 101
2   N-body dynamics 101
3   Invariants, approximate motion and collisions 102
4   Collisions and Lyapunov exponents 104
5   Kinetic theory and BBGKY hierarchy............................... 105
6   Mean-field limit and Vlasov equation 107
7   Vlasov–Poisson equation for Coulomb and Newton interactions....... 109
8   Boltzmann–Grad limit and Boltzmann equation 110
9   Entropy dissipation for the Boltzmann equation 111
References 113

Chaotic Scattering in Planetary Rings
Jean-Marc Petit 114
1   Introduction 114
2   Dynamics of planetary rings 115
     2.1   The physical problem 115
     2.2   Equations of motion 115
     2.3   Chaotic scattering 122
     2.4   Other examples... 125
3   Symbolic dynamics 127
     3.1   The Bernoulli shift.. 129
     3.2   Topological mappings 130
     3.3 C^1 mappings 131
     3.4   Homoclinic points.. 132
4   The inclined billiard 132
     4.1   The model and an interesting limit......................... 133
     4.2   Properties of the motion 137
     4.3   Symbolic dynamics.. 141
References 144

Close Encounters in Öpik's Theory
Giovanni B. Valsecchi 145
1   Introduction 145
2   Basic formulae of Öpik's theory 146
     2.1   The components of the planetocentric velocity 146
     2.2   The angles θ and ϕ 147
     2.3   The rotation of U 148
3   From the planetocentric to the b-plane frame and back 151
     3.1   The ecliptic on the b-plane 152
     3.2   The projection of the X-axis on the b-plane 152
     3.3   The projection of the Y-axis on the b-plane 152
     3.4   The projection of the Z-axis on the b-plane 153
     3.5   Examples 153
4   The motion of the small body...................................... 154

4.1   The local Minimum Orbital Intersection Distance (MOID) 155
4.2   The planetocentric orbital elements of the small body 156
4.3   The encounter 157
4.4   The new local MOID 160
4.5   Post-encounter coordinates in the post-encounter b-plane 161
4.6   The next encounter 162
5   Resonant returns in Öpik's theory 164
5.1   Solving for a given final semimajor axis 164
5.2   Examples 165
6   The distribution of energy perturbations 166
6.1   Energy perturbations for a given MOID 170
7   Geocentric variables to characterize meteor orbits 170
7.1   An orbital similarity criterion based on geocentric quantities 171
7.2   Secular invariance of U and θ 175
References 177

Generalized Averaging Principle and Proper Elements for NEAs
Giovanni-Federico Gronchi 179
1   Introduction 179
2   The classical averaging principle 180
2.1   The full equations of motion 180
2.2   The averaged equations 181
2.3   Difficulties arising with crossing orbits 183
3   Generalized averaging principle in the circular coplanar case 183
3.1   Geometry of the node crossing 183
3.2   Description of the osculating orbits 186
3.3   Weak averaged solutions 187
3.4   The Wetherill function 188
3.5   Kantorovich's method 190
3.6   Integration of $1/d$ 191
3.7   Boundedness of the remainder function 194
3.8   The derivatives of the averaged perturbing function $\overline{R}$ 196
4   Secular evolution theory 199
4.1   The secular evolution algorithm 201
4.2   Different dynamical behavior of NEAs 202
5   Proper elements for NEAs 203
6   Reliability tests 207
7   Generalized averaging principle in the eccentric–inclined case 208
7.1   The mutual reference frame 209
8   Conclusions 210
References 210

Subject Index 213

List of Contributors

Benest, Daniel
O.C.A. Observatoire de Nice
B.P. 4229
F-06304 Nice Cedex 4, France
benest@obs-nice.fr

Celletti, Alessandra
Dipartimento di Matematica,
Università di Roma "Tor Vergata"
Via della Ricerca Scientifica 1
I-00133 Roma, Italy
celletti@mat.uniroma2.it

Della Penna, Maria Gabriella
O.C.A. Observatoire de Nice
B.P. 4229
F-06304 Nice Cedex 4, France
now at
Via Istonia n.47
I-66051 Cupello (ch), Italy
mgdellapenna@libero.it

Elskens, Yves
Equipe turbulence plasma,
Laboratoire de physique des interactions ioniques et moléculaires
case 321, Campus Saint-Jérôme
F-13397 Marseille Cedex 20, France
elskens@newsup.univ-mrs.fr

Falcolini, Corrado
Dipartimento di Matematica,
Università di Roma "Tor Vergata"
Via della Ricerca Scientifica 1
I-00133 Roma, Italy
falcolin@mat.uniroma2.it

Froeschlé, Claude
O.C.A. Observatoire de Nice
B.P. 4229
F-06304 Nice Cedex 4, France
claude@obs-nice.fr

Gronchi, Giovanni-Federico
Dipartimento di Matematica,
Università di Pisa
Via Buonarroti 2
I-56127 Pisa, Italy
gronchi@newton.dm.unipi.it

Petit, Jean-Marc
O.C.A. Observatoire de Nice
B.P. 4229
F-06304 Nice Cedex 4, France
now at
Observatoire de Besançon
41 bis Avenue de l'Observatoire
F-25010 Besançon, France
petit@obs-besancon.fr

Valsecchi, Giovanni
Istituto di Astrofisica Spaziale,
Area di Ricerca del C.N.R.
via Fosso del Cavaliere 100
I-00133 Roma, Italy
giovanni@ias.rm.cnr.it

Waldvogel, Jörg
Applied Mathematics,
ETH
CH-8092 Zurich, Switzerland
waldvoge@math.ethz.ch

Singularities, Collisions and Regularization Theory

Alessandra Celletti

Dipartimento di Matematica, Università di Roma "Tor Vergata", Via della Ricerca Scientifica, 1 - I–00133 Roma (Italy)

Abstract. An overview of singularities, regularizations and collisions in gravitational N–body systems is presented. The concept of singularity pertains to many fields of science, from the Big Bang theory to black holes, atomic physics, etc. As far as gravitational critical phenomena are concerned, a plethora of collisional events marked the history and evolution of the solar system. The Earth itself experienced many collisions from prehistoric age to recent times due to impacts of asteroids or comets. We report several examples of meteorites and we provide the rate of an impact as a function of the diameter of the colliding object. The standard classification of Near–Earth Objects is presented. From the theoretical point of view, the singularity due to binary collisions between point masses can be handled by means of regularization theory. We review this technique for the limiting case of a two–body system on a line. Coordinate transformations, the introduction of a fictitious time and the conservation of the energy are used to regularize the equations of motion. Triple collisions and the concept of the central manifold are discussed. A simple model, known as the inclined billiard, is presented to investigate chaotic diffusion. Symbolic dynamics is used to characterize the motion, which closely resembles the trajectory of a ring particle. The problem of noncollision singularities is discussed from Painlevé's conjecture to a 5–body example of noncollision singularities.

1 Introduction

In mathematics and physics, a singularity denotes an anomalous event. It is not usually welcomed due to the fact that it represents the point at which theories become more complicated. Quoting G.L. Schroeder ([19]), "It is essential to bear in mind that science has not provided explanations for the two principal starting points in our lives: the start of our universe and the start of life itself". The transition instant between nothingness and the beginning of the universe or of life is a singular event.

Singularities occur in many fields of science, from mathematical analysis to cosmology through the theory of the Big Bang and the concept of black holes. We review some of the most important examples involving singular events in Sect. 2. However, the main purpose of this work is to present an introduction to singularities in the framework of a *gravitational* system of point masses. From Newton's law, we know that bodies interact by means of a force which is proportional to the inverse of the squared distance. As the two bodies approach each other, their distance tends to zero and, consequently, the differential equation describing the dynamics of the system becomes singular when the two bodies collide. In the real world the point masses are replaced by bodies of finite (non zero) size; in this case, we refer to a binary collision whenever the distance between their centers of mass equals the sum of the radii. The history of the solar system has been largely marked by such catastrophic events. The widely cratered surfaces of rocky satellites or planets indicate that collisions were very frequent in the past, particularly in the early stages of the formation of the solar system. The Earth itself experienced collisions in different epochs, from prehistoric age to recent times. Nowadays, the disappearance of dinosaurs is widely attributed to a collisional event. The occurrence of impacts with the Earth represents a concrete possibility, due to the large number of asteroids or comets which intersect the orbit of the Earth. Those bodies which come close to our planet are denoted as "Near–Earth Objects" (NEOs). We report in Sect. 3 several examples of past impacts, the classification of NEOs and a discussion of the probability of collisions.

Even in the simplified approach that solar system objects are reduced to point masses, the description of the dynamics of an N–body system becomes difficult during close encounters, due to the loss of regularity of the equations of motion. In order to get rid of this problem, regularization theories ([10], [20], [21], [22], [24], [26]) have been developed to transform a singular equation into a regular one. We mainly focus in Sect. 4 on the so–called Levi–Civita regularization, which is based on an *ad hoc* transformation of the spatial variables and of time (through the introduction of a fictitious time), taking advantage of the conservation of energy. In order to familiarize with this technique, we investigate a simple model problem consisting of two bodies on a line (i.e., in the limit of orbital eccentricity equal to one). More realistic models involving triple collisions ([29], [30]) are reviewed in Sect. 5, where the concept of *central configurations* and *homographic solutions* is introduced. A simple model, known as the *inclined billiard*, showing the main features of chaotic scattering in N–body dynamics, has been introduced by M. Hénon ([8], [9], [17]) and will be presented in Sect. 6.

Collisions are not the only cause of singularities in gravitational systems of N point masses. As P. Painlevé conjectured, there might exist motions leading to singularities without experiencing collisions. More precisely, from a result of H. von Zeipel ([31]) noncollision singularities might occur if the dynamics becomes unbounded in a finite time. Though Painlevé himself proved that in the framework of the 3–body problem the only possible singularities are collisions, the question remains open as far as N–body systems with $N \geq 4$ are investigated. This conjecture remained open for about a century, until Z. Xia ([27]) provided an analytical proof of the existence of noncollision singularities in a 5–body model. Suitable occurrences of close–triple encounters are the main ingredient of the proof. However, we stress that until nowadays there is no proof of Painlevé's conjecture in the context of the 4–body problem. Noncollision singularities are discussed in Sect. 7.

Let us conclude with the following (possibly optimistic) remark: collisions are definitely dangerous dramatic events. But if we accept the theory that dinosaurs disappeared due to the impact of a large body with the Earth, we can state that mankind could not have evolved without such a tragic event which provoked a long night lasting more than one year.

> *There is no Sun without shadow,*
> *and it is essential to know the night.*
> (Albert Camus)

2 A world of singularities

The concept of singularity does not pertain only to gravitational systems, but it extends also to many other fields of science, from the atomic structure of matter to cosmology. In this section we provide some examples of singularities which arise in different contexts.

From the mathematical point of view, singular points play a fundamental role in the study of functions of complex variables, since their nature (pole, branch point, essential singularity) and their distribution characterize the function. Similarly, nodes, cusps or isolated points are relevant in the study of algebraic curves, i.e. the set of points satisfying a polynomial equation. We do not want to insist too much on the mathematical aspect (for which the existing literature is extremely wide), but rather point to several physical examples having a singularity as common feature. The idea of a singularity is often related to the *Big Bang* theory, according to which the origin of the universe was a singular point. All time, space, energy and matter were concentrated in a point with infinite density. The universe exploded from the initial singularity up to the present state, whereby galaxies, stars and planets formed. The future of the universe crucially depends on its total mass: it might undergo an infinite expansion (open universe), it might reach a stationary state or it might expand until a subsequent contraction takes place. In the last case, one is faced with a new singularity at the *Big Crunch*. Experimental evidence in support of the Big Bang theory, like the detection of the background microwave radiation, dates back to the second half of the last century. However, it is worth mentioning that a cosmological model based on the birth of the universe from a singularity was already present in Stoic philosophy, denoted with the term *ekpyrosis*, which means conflagration. Moreover, Greeks already imagined that a Big Bang process might indefinitely repeat itself.

General relativity is at the basis of the Big Bang model. Einstein's theory provides a description of the gravitational field surrounding a massive body. The spacetime is curved and the curvature stretches matter in all directions. Einstein's solutions can admit a singular point at which the curvature of spacetime is infinite, the simplest being the *Schwarzschild* solution. This solution is the starting point of the notion of *black holes*, whose center is a singularity surrounded by a surface that hides its content from the visible universe. The limiting surface is called the *horizon of events*. Although we do not have direct evidence of the existence of black holes, Hubble Space Telescope images and Chandra X–ray observations provided many plausible candidates. These anomalies of the universe originate from very massive stars which terminate in a supernova explosion. Their gravitational field is so strong, that light cannot escape from them. Black hole's singularities came recently into the scene as the entrance key toward other universes, through a spatiotemporal wormhole connecting two black holes in different universes.

From the Big Bang and black holes, singularities bridge to the infinitely small in connection to the atomic structure of the universe. For example, collisions of matter and antimatter produce quarks through annihilation. On the same scale, chemical reactions are founded upon atomic or molecular collisions, whose frequency determines the velocity of the reaction. In biology, the processes of meiosis and mitosis can also be viewed as critical events with different conclusions: in the first case the genetic component is equally distributed between the cells produced by the fission, while in the process of mitosis the new cells retain the same genetic inheritance as the original one. Let us mention that a gravitational approach has been also adopted to detect neuronal groups identified by temporally related firing patterns ([5], [11]). However, singularities are not welcomed in this case; to this end, a minimal distance between neurons is introduced, such that the net force is zero. This technique to avoid singularities is the same as that used to integrate many–body systems, like a large number of stars belonging to the same galaxy.

3   Past and future collisions in the Solar System

The problem of collisions in our planetary system is far from being a theoretical topic. The cratered surfaces of many planets and satellites are the imprints of dramatic events that happened in the early stages of the formation of the solar system. However, catastrophic encounters belong also to the recent history of the solar system and definitely they will influence the future evolution of planets and satellites. Collisions have determined the development of Earth, as the disappearance of dinosaurs is almost certainly due to an impact of a heavy object, and might (hopefully not!) influence the future of human lives. These events might be due to asteroids, comets or meteorites. To be precise, let us give a qualitative definition of the candidate impact bodies.

• *Asteroids* are rocky bodies of relatively small size (typically some tens to hundreds of kilometres) which orbit around the Sun. They form the so–called *asteroidal belt* between the orbits of Mars and Jupiter. Collisions between asteroids are rather frequent. They can be pulled into the inner solar system by resonant interactions or by the gravitational influence of Jupiter.

• *Comets* are relatively small bodies composed mainly of ice, that vaporizes in the proximity of the Sun giving rise to a tail of dust and gas. Orbits of comets can be elliptic, parabolic or hyperbolic. It is widely accepted that a reservoir of comets is the *Oort cloud*, a region far outside the solar system at a distance of about $10^4 - 10^5$ AU (1 AU is the average distance of the Earth from the Sun).

Comets are usually divided into long–period (on highly eccentric orbits) and short–period (less than 200 years) comets.

• *Edgeworth-Kuiper objects:* news from the edge of the solar system. A new group of icy trans-neptunian objects with diameters of some hundreds kilometres has been recently observed at a distance of about 35–50 AU.

• *Meteoroids* are small particles originating from a comet or an asteroid which orbits the Sun.

• *Meteors* are the light phenomena due to the interaction of a particle with the Earth's atmosphere, giving rise to so–called *shooting stars.*

• *Meteorites* are bodies sufficiently large to survive the impact with the Earth's atmosphere and to land upon the surface of our planet.

The study of solar system objects coming close to the Earth is particularly important for several reasons: 1) the determination of their physical and chemical composition suggests a possible scenario for the origin and evolution of the solar system; 2) these objects are possible reservoirs of raw materials; 3) they should be monitored to control the risk of collisions with the Earth.

Concerning the last point, observations show that almost one hundred tons of interplanetary matter hit the surface of the Earth every day. Most of the smaller particles are dust grains. Moreover, recent estimates suggest that every year almost 500 meteorites of at least 100 g arrive on every million km^2 of the Earth's surface. However, the number of meteorites collected is much smaller.

Impacts of objects of about 50 metres diameter with the Earth occur about every 100 years in the average. As an example, the *Meteor Crater* of northern Arizona formed 50 000 years ago, when an asteroid of about 30 m diameter fell on the Earth. The crater is about 1.2 km wide and 170 m deep. There are many astroblames on the Earth provoked by impacts of Near–Earth Objects (NEO's). Let us quote another example which puzzles scientists since 30 June 1908. On that day, at 7:17 a.m., an enormous fireball crossed the sky over the Tunguska region in Siberia. Over 2 000 km^2 of the nearby forest were destroyed, and anomalous earthquakes were detected even at 800 km of distance. A shock wave reached England after about 5 hours and returned back to Siberia after 24 hours. However, since no remains of the responsible were ever found, scientists advanced the hypothesis that the colliding body was a comet evaporated in the atmosphere. More recently, it was conjectured that the killer object was a rocky asteroid of about 60 metres diameter, which exploded and disintegrated 5–10 km from the Earth's surface.

The impact with the Earth of a NEO of about 1 km in diameter is evaluated to occur every few hundred thousand years in the average. The energy released during the impact of a rocky object of that size can be computed evaluating the kinetic energy as $T = \frac{1}{2}mv^2$, where one can assume that the density is about 3 g/cm^3 and the typical velocity is about 20 km/s. Such an estimate leads to the prediction that the energy released would be of the order of magnitude of some million nuclear explosions. A larger object (say, 10–20 km diameter) would of course threaten life on Earth, provoking a climatic and biological disaster on planetary scale, like the one which caused the disappearance of the dinosaurs. The evidence that this event was generated by an asteroidal collision dating back 65 million years was recently attributed (thanks to the images of the shuttle *Endeavour*) to the discovery of a large crater (about 180 km) in the depth of the ocean close to the Yucatan peninsula. In honour of the Maya civilization, the crater was named "Chicxulub", the devil's tail. The abrupt fall in temperature, the decrease of the sea level and the transformation of the vegetation caused the extinction of about 65–75% of the living species. A similar event dating back 210 millions years and testified by the Manicouagan crater in Quebec might have been the cause of a mass extinction of marine species at the end of the Triassic period.

Concerning other bodies of the solar system, astroblames are evident from any picture of rocky planets or satellites. In 1994, we were extremely lucky to witness the spectacular collision of the Shoemaker–Levy 9 comet with Jupiter. The comet fragmented in several pieces before the impact, due to the gravitational influence of the giant planet. The first fragment, which travelled at a speed of about 200000 km/h caused the emission of hot matter to a distance of 3000 km from Jupiter's surface. Finally, let us mention that one of the most important theories on the formation of the Moon takes into account the possibility that our satellite formed from a catastrophic impact of a Mars–size object with the Earth. Lunar rocket samples collected during the Apollos missions support the hypothesis that a big body crashed into the Earth during its early formation stages, provoking the ejection of material which, after coagulation, gave origin to our Moon.

Coming back to the Earth, the most dangerous objects are those which approach our planet during their travel in the solar system. For example, the asteroid 4179 Toutatis of 2–3 km of diameter is one of the most alarming ones, since every 4 years it crosses the Earth's orbit. In December 1992, 4179 Toutatis was at a distance of 3.6 million kilometres and it will come as close as 1.5 million kilometres in September 2004. We can collect the above examples providing an estimate of the probability of an impact as a function of the diameter of the colliding object (see Table 1). However, we have to bear in mind that we cannot rely on probability to exclude the possibility of an impact on Earth in the near future. As an example, we mention that in 1971 the roof of a house in Wethersfield (Connecticut) was damaged by an object of a mass of 340 g. Eleven years later, an object of 2.5 kg destroyed another roof in the same town!

We report in Table 2 the characteristics of the most important craters found on Earth ([16]). In particular we provide the diameter of the main crater (since the impact can generate more than one crater) and the year of the discovery.

The biggest meteorite discovered on the Earth is still in the place where it arrived in prehistoric times, namely in Hoba West (Southwest Africa). Its weight amounts to about 61 tons. A list of the biggest meteorites is given in Table 3 ([16]).

Table 1. Impact frequency as a function of the diameter d of the colliding body.

d (km)	Frequency (years)
$d > 10$	50 000 000
$1 < d < 10$	500 000
$0.1 < d < 1$	5 000
$0.03 < d < 0.1$	500

Table 2. Most relevant craters on Earth: the diameter refers to the main crater.

Name		Diameter (m)	Year of discovery
Meteor Crater,	Arizona	1 265	1891
Wolf Creek,	Australia	850	1947
Henbury,	Australia	200×110	1931
Boxhole,	Australia	175	1937
Odessa,	Texas	170	1921
Waqar,	Arabia	100	1932
Oesel,	Estonia	100	1927
Campo del Cielo,	Argentina	75	1933
Dalgaranga,	Australia	70	1928
Sichote–Alin,	Siberia	28	1947

Table 3. Biggest meteorites and place of discovery.

Name	Place of discovery	Weight (tons)
Hoba–West	Africa	61
Ahnighito	Greenland	30.9
Bacuberito	Mexico	27.4
Mbosi	Tanganyika	26.4
Agpalik	Greenland	20.4
Armanty	Mongolia	20
Willamette	USA	14
Chupaderos	Mexico	14
Campo del Cielo	Argentina	13
Mundrabilla	Australia	12
Morito	Mexico	11

Near–Earth Objects are commonly defined as asteroids or comets, such that their perihelion distance q is less than 1.3 AU. Moreover, denoting by a the semimajor axis, by Q the aphelion distance and by P the orbital period, one has the following classification:

• Near–Earth Comets (NECs): short–period comets with $P < 200$ years and $q < 1.3$ AU.

• Near–Earth Asteroids (NEAs): asteroids with perihelion distance $q < 1.3$ AU. The NEAs are further classified as

  • Atens (named after asteroid 2062 Aten) are Earth–crossing asteroids with $a < 1$ AU and $Q > 0.983$ AU;

  • Apollos (named after asteroid 1862 Apollo) are Earth–crossing asteroids with $a > 1$ AU and $q < 1.017$ AU;

  • Amors (named after asteroid 1221 Amor) are Earth approaching asteroids with $a > 1$ AU and $1.017 < q < 1.3$ AU.

• Potentially Hazardous Asteroids (PHAs): NEAs with Minimum Orbit Intersection Distance (MOID) with the Earth less than or equal to 0.05 AU and with an absolute magnitude less than or equal to 22.

We remark that the MOID introduced in the definition of PHAs is a measure of the possibility that an asteroid may make alarming close approaches with the Earth. In other words, it provides the local minimum of the distance of the NEO to the Earth. Using suitable variables introduced by the astronomer E. Öpik, an approximate expression for the local MOID at a specific node can be derived (see the chapter by G.B. Valsecchi, in this book). The secular evolution of the orbits of NEOs can be approached through an appropriate *averaging principle* as presented in the chapter by G.F. Gronchi, in this book. Let us mention that another measure of the risk of collisions of asteroids or comets with the Earth is the *Torino impact scale*, which ranges from zero (implying no real damages) to ten (involving a climate catastrophe). The impact probability can be calculated on the basis of experimental observations or theoretical analyses. As an example, we report in Table 4 some asteroids with possible impact solutions which are consistent with recent observations. The data are taken from the web site of NEODyS (http://newton.dm.unipi.it/neodys/).

The trajectory of a NEO can be determined statistically, within a region of confidence. However, the semimajor axis, the eccentricity and the inclination of the orbit vary only on a long timescale and can be considered as almost fixed over a short time. Therefore, it turns out to be useful to group the objects within different dynamical classes. This task is quite difficult as far as planet–crossing orbits are concerned, due to their high degree of chaoticity and due to the many close encounters with one (or more) planets. However, from the mean evolution of the orbits, one can derive a probability of collision as it was suggested by Öpik. The conditions required for an impact between a NEO and the Earth to occur are the following:

Table 4. Asteroids with possible impact solutions with the Earth: a denotes the semi-major axis, e the orbital eccentricity, i the inclination, H the absolute magnitude and the MOID is referred to the Earth.

Name	a(AU)	e	$i(^\circ)$	H	$MOID$(AU)
1994GK	1.9915	0.6122	5.723	24.197	0.00292
1994UG	1.2293	0.2581	4.682	21.134	0.01171
1994WR12	0.7566	0.3979	6.856	22.107	0.00188
1995CS	1.8732	0.7643	2.556	25.501	0.00134
1997TC25	2.5341	0.6136	0.247	24.665	0.00132
1998OX4	1.5857	0.4878	4.547	21.328	0.00148
2001AV43	1.2771	0.2381	0.279	24.322	0.00154
2001SB170	1.3457	0.4580	34.148	22.145	0.00413
2001TY1	2.4117	0.5890	5.820	24.856	0.00488
2001UD5	2.2751	0.6654	2.530	22.205	0.00266

i) the orbits of the Earth and of the NEO must cross, even if the mutual inclination is different from zero;

ii) the nodal distance between the two orbits must be smaller than the dimension of the Earth;

iii) the NEO and the Earth must arrive at the same time at the intersection between the two orbits.

The computation of the probability of impact, the monitoring of the known NEOs and the search for new possible dangerous objects is a primary goal which brings together astronomers, mathematicians, physicists, etc. As geological history (from prehistoric age to recent times) showed us, impact events are far from being unusual. To this end, several groups of scientists are developing projects to make Earth safer. Let us mention, for example, the works of the NASA, Jet Propulsion Laboratory, Spaceguard Foundation, Space Mechanics Group of the University of Pisa and the Minor Planet Center of the International Astronomical Union.

4   Regularization theory

Throughout the remaining sections we consider N–body systems composed by N point masses. The occurrence of a collision is a very difficult subject to handle from the mathematical point of view, since the equations of motion cease to be valid at the singularity. As a consequence of the conservation of the energy, since the potential function is infinite at collision, the velocity becomes itself infinite. The description of the motion fails at the singularity, but what is even worse, it is rather difficult to investigate the dynamics in a *neighbourhood* of the singularity. Even excluding a collision, it is anyway troublesome to explore the trajectories corresponding to close approaches: the integration step required in a numerical approach is usually very small, thus requiring a huge amount of computational time.

A way to overcome these difficulties has been explored by several mathematicians at the end of the XIX century and at the beginning of the XX century. Among others, T. Levi–Civita, G.D. Birkhoff, P. Kustaanheimo, E.L. Stiefel, K.F. Sundman, C.L. Siegel, J.K. Moser, J. Waldvogel ([10], [20], [21], [22], [23], [24], [26], [28]) contributed to develop a theory of regularization for the study of the motion at a collision. Planar and spatial three–body problems have been investigated, as well as the occurrence of binary or triple collisions. At the beginning of the last century, K.F. Sundman was able to prove that a solution which does not experience a collision can be expanded as a power series, which converges at any time ([24]). In the case of a binary collision at time $t = t_c$, he proved that the solution can be written as a convergent power series in terms of $(t_c - t)^{1/3}$. It is impossible to holomorphically extend the solution up to t_c, but Sundman found a real analytic continuation for $t > t_c$, using complex analytic continuation around t_c. Let us mention that a general definition of regularization was given by R. Easton ([4]), who developed the so–called *block regularization* in order to investigate whether neaby orbits provide an extension for an orbit ending into a collision. This procedure of pasting orbits is denoted as *Easton's method*.

Instead of using Sundman's or Easton's approach, we will rather be concerned with the Levi–Civita regularization, which is based upon three main steps:

• the introduction of a suitable change of coordinates (typically the *Levi–Civita transformation*);

• a stretching of the time scale, to get rid of the fact that the velocity becomes infinite (namely, the introduction of a so–called *fictitious time*);

• the conservation of the energy, to transform the singular differential equations into regular ones (i.e., the study of the Hamiltonian system in the *extended phase space*).

It is important to bear in mind that the aim of regularization theory is to transform singular differential equations into regular ones. As an example, consider the second order differential equation

$$\ddot{x}\,(1 - \sin t) + \dot{x} - x\,\sin t + 2\,\sin t = 0\,,$$

which can be written as

$$\ddot{x} = \frac{1}{1 - \sin t}\,(-\dot{x} + x \sin t - 2 \sin t)\,.$$

The equation is singular for $t = \frac{\pi}{2}$ mod 2π, but it has a regular solution,

$$x(t) = \sin t + \cos t\,.$$

Since the description of the regularizing transformations is technically complicated in the general setting of the three–body problem, we propose a preliminary presentation of such theory using a very elementary (physically trivial) model problem. More precisely, we start by considering the problem of two bodies moving under their mutual gravitational attraction. In analogy to [22], we further simplify our task by assuming that the two bodies move on a straight line. Let us retrace the above mentioned steps necessary for regularization, looking at the effect of each transformation on the dynamics of the colliding object. Moreover, at each step we devote particular attention to the case when the energy is zero.

Let P_1 and P_2 be two bodies with masses m_1 and m_2. We fix a reference frame with the axes parallel to an inertial system and with the origin coinciding with the body P_2. We restrict P_1 to move on the x–axis. Let K be the product of the gravitational constant G with the sum of the masses, i.e. $K = G(m_1 + m_2)$. Then, the motion of P_1 with respect to P_2 is governed by the differential equation

$$\ddot{x} + \frac{K}{x^2} = 0 \, , \tag{1}$$

which admits the integral of energy

$$h = \frac{K}{x} - \frac{1}{2}\dot{x}^2 \, .$$

Denoting the velocity by $y = \dot{x}$, one obtains $y = \pm\sqrt{2(\frac{K}{x} - h)}$, showing that it becomes infinite at the collision ($x = 0$) and it is zero for $x = \frac{K}{h}$. If the initial velocity is positive, then P_1 gets indefinitely far away from P_2. On the other hand, if the initial velocity is negative the collapse of P_1 on P_2 is unavoidable. In Fig. 1 we show the integration of the equation of motion by a 4th order Runge–Kutta method with time step equal to 10^{-3}, for $h = 0.5$ and the initial position $x_0 = 1$ (K has been normalized to one). The graphs of the variation of the position with the time and the phase space diagram are presented for a) positive initial velocity and b) negative initial velocity.

Let us now consider the case $h = 0$. From $\dot{x} = \pm\sqrt{\frac{2K}{x}}$, the equation of motion can be analytically integrated as

$$x(t) = \left[x_0^{3/2} \pm \frac{3}{2}\sqrt{2K}\, t\right]^{2/3} , \tag{2}$$

where $x_0 > 0$ denotes the initial position at time $t = 0$. The plus or minus sign has to be taken according to whether the initial velocity is positive or negative; in the latter case the solution is valid up to the collisional time, say $t = t_c$. The graph in Fig. 2 shows that the solution with positive initial velocity indefinitely departs, while the solution with negative initial velocity tends to the singularity $x = 0$. The aim of the regularization procedure is the analogue of removing a branch point in the field of complex variables, which is a standard technique known as *uniformization*.

• Change of coordinates:
In the same spirit of the regularizing transformations for the general case, we

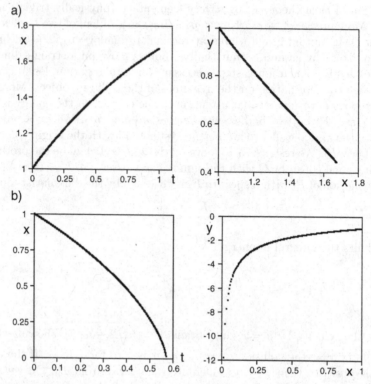

Fig. 1. Integration of Equation (1) for $h = 0.5$, $x_0 = 1$. *a)* Positive initial velocity; *b)* negative initial velocity.

perform a change of coordinates, replacing x by

$$x = u^2 \ .$$

Although this transformation is not really necessary to regularize the motion of the two–body problem on a line, we prefer to include this step in our discussion, since its generalization to more–dimensional spatial coordinates in the framework of the three–body problem is an essential ingredient, known as the *Levi–Civita transformation*. From (1), the equation of motion in the u–variable becomes

$$\ddot{u} + \frac{1}{u}\dot{u}^2 + \frac{K}{2u^5} = 0 \ , \tag{3}$$

while the energy equation is

$$h = \frac{K}{u^2} - 2u^2\dot{u}^2 \ ,$$

from which there follows that

$$\dot{u} = \pm\sqrt{\frac{K}{2u^4} - \frac{h}{2u^2}} \ .$$

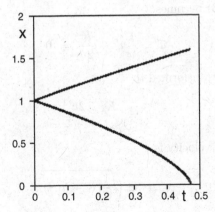

Fig. 2. The solution (2) for $h = 0$, $x_0 = 1$. The portion of the curve above $x = 1$ corresponds to taking the positive sign in (2); the portion below $x = 1$ corresponds to the negative sign in (2).

We immediately recognize that the equation is still singular and that the velocity is infinite at the singularity ($u = 0$). The graph of u as a function of the time, as well as that of u versus $\dot{u}$, are very similar to those previously shown in Fig. 1.

In the case $h = 0$, the solution can be given explicitly as

$$u(t) = \left[u_0^3 \pm 3\sqrt{\frac{K}{2}}\, t\right]^{1/3},$$

where $u_0 = u(0)$ represents the initial condition at $t = 0$ and we adopt the plus or minus sign according to the direction of the velocity.

• Introduction of the fictitious time:
In order to get rid of the increase of the speed to infinity at collision, we multiply the velocity by an appropriate scaling factor which is zero at the singularity. In other words, we introduce a *fictitious time s* defined by

$$dt = x\, ds = u^2\, ds \quad \text{or} \quad \frac{dt}{ds} = x = u^2, \tag{4}$$

namely

$$s - s_0 = \int_{t_0}^{t} \frac{1}{x(\tau)}\, d\tau$$

(see (7) below). Then, by introducing the notation $u' \equiv \frac{du}{ds}$, one has

$$\dot{u} = \frac{du}{dt} = \frac{1}{u^2} u',$$

$$\ddot{u} = \frac{1}{u^2}\frac{d}{ds}\left(\frac{1}{u^2} u'\right) = \frac{1}{u^4} u'' - \frac{2}{u^5} u'^2.$$

The equivalent of (3) becomes

$$u'' - \frac{1}{u}u'^2 + \frac{K}{2u} = 0 , \tag{5}$$

while the energy is transformed to

$$h = \frac{K}{u^2} - \frac{2u'^2}{u^2} , \tag{6}$$

from which the new velocity is

$$u' = \pm\sqrt{\frac{1}{2}(K - hu^2)} .$$

The solution for $h = 0$ can be easily found from $u' = \pm\sqrt{\frac{1}{2}K}$, namely

$$u(s) = u_0 \pm \sqrt{\frac{1}{2}K}\, s ,$$

where u_0 is the initial value of u at $s = 0$, while (here and in the following formula) the plus or minus sign must be chosen in accordance with the sign of u'. Furthermore, from (4) we obtain the dependence of the time t upon the fictitious time s as

$$t = u_0^2 s + \frac{K}{6}s^3 \pm \sqrt{\frac{1}{2}K}\, u_0\, s^2 . \tag{7}$$

The graph of $t = t(s)$ (with $u_0 = 1$) is presented in Fig. 3.

• Using the energy integral:
Finally, we use the fact that the energy is preserved by inserting (6) in (5) to obtain

$$u'' + \frac{h}{2}\, u = 0 ,$$

which is the equation of the harmonic oscillator with frequency $\omega = \sqrt{\frac{h}{2}}$. We therefore have succeeded in obtaining a regular differential equation, whose solution is a periodic function of s. In the elliptic problem, it can be shown that the fictitious time s is essentially the eccentric anomaly.

The regularization of the two–body problem on a line contains all the main ingredients to perform regularization of more sophisticated (and realistic) problems. Next step is to consider the simplest three–body model, i.e. the planar, circular, restricted three–body problem. Denoting by P_1, P_2, P_3 three bodies with masses m_1, m_2, m_3, we assume that

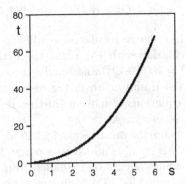

Fig. 3. Graph of $t = t(s)$ as in (7) for $h = 0$, $x_0 = 1$ after the introduction of the fictitious time.

i) the mass of one particle, say P_3, is much smaller than the others ("restricted" problem) so that it does not influence the motion of the primaries P_1 and P_2;
ii) the motion of the three bodies takes place in the same plane ("planar" problem);
iii) the relative motion of the primaries is a circle ("circular" problem).

It is convenient to write the equations of motion of P_3 under the gravitational influence of P_1 and P_2 in a synodic reference frame, which rotates with the angular velocity of the primaries. The classical setting puts the origin at the center of mass, and the fixed positions of P_1 and P_2 are on the axis of the abscissas. Let us denote by (q_1, q_2) the coordinates of P_3 in the synodic frame and by (p_1, p_2) the corresponding momenta. The motion of P_3 is governed by the Hamiltonian

$$H(p_1, p_2, q_1, q_2) = \frac{1}{2}(p_1^2 + p_2^2) + q_2 p_1 - q_1 p_2 - \frac{m_1}{r_1} + \frac{m_2}{r_2},$$

where m_1, m_2 are the masses of P_1, P_2 and r_1, r_2 denote the distances of P_3 from P_1 and P_2, respectively. A singularity occurs when P_3 collides with P_1 (i.e., $r_1 = 0$) or with P_2 (i.e., $r_2 = 0$). The first step is therefore to regularize the collision with one of the primaries. This task is achieved by retracing the technique introduced for the rectilinear two–body problem. The change of coordinates must be replaced now by the transformation introduced by Tullio Levi–Civita (born in Padua 1873, died in Rome 1941). Like many other mathematicians of his time, Levi–Civita contributed to many fields of mathematics, from differential geometry to analysis, relativity and mechanics. A crater of our Moon is named after him. The transformation which bears his name is the first tool to regularize the equations of motion and takes the form:

$$q_1 = Q_1^2 - Q_2^2 + \alpha \,, \qquad q_2 = 2Q_1 Q_2 \,,$$

where $\alpha = m_2$ is used in order to regularize a collision with P_1, while $\alpha = -m_1$ handles the case of a collision with P_2. Then, the fictitious time is introduced by $dt = D\,ds$, where $D = 4(Q_1^2 + Q_2^2)$ and finally, the conservation of the energy (i.e., the definition of the Hamiltonian in the extended phase space) allows to obtain the regularized equations of motion (further details can be found in the chapter by A. Celletti, in this book).

A slightly different technique must be used when we allow the three bodies to move in space, rather than being confined to a plane. More precisely, the Levi–Civita transformation cannot easily be extended from 2 to 3 dimensions. The physical space must be embedded in a 4–dimensional space, by introducing an extra variable. The study of the collisions in the spatial case can be performed by means of the so–called Kustaanheimo-Stiefel regularization ([10]). An alternative technique to regularize the equations in the planar and spatial cases is provided by the radial–inversion transformation developed in [20].

There still remains the task of simultaneously regularizing both collisions with the primaries. This problem was solved in [1] for the planar case and in [23], [28] for the spatial case. We refer to the chapter by G. Della Penna, in this book, for further details, where the problem is studied again in the framework of the planar, circular, restricted three–body problem.

All regularizing procedures outlined before required that the time is replaced by a fictitious time. However, it is worth mentioning that a different approach can be adopted substituting the time with the generalized eccentric anomaly. This procedure turns out to be efficient while dealing with very elongated orbits. A comparison between the two methods, as well as an application of perturbation techniques (to integrate the equations to first order), is presented in the chapter by C. Falcolini, in this book.

We have overviewed the regularization for the 2 and 3–body problems. However, the simulation of more general systems, like the motion of stars in a galaxy, requires the study of N–body problems with large N. In this case, it is difficult to apply the above techniques, and a way to overcome this hindrance is to apply Plummer's method ([3], [15]). A positive softening length is introduced to eliminate the singularity. More precisely, one replaces the gravitational force on the j–th body $(1 \leq j \leq N)$ by Plummer's force defined as

$$F_j \equiv \sum_{i=1}^{N} \frac{x_j - x_i}{(|x_j - x_i|^2 + \varepsilon^2)^{3/2}} \,, \tag{8}$$

where x_i denotes the barycentric Cartesian position of the i–th particle, and the softening length ε defines the degree of smoothing. One immediately recognizes that the function (8) does not lead to singularities at the collision and can be used in an effective way in numerical simulations of large N–body systems ([3], [15]).

When the number of bodies becomes infinite, namely in the limit when $N \to \infty$, a different approach must be used in the framework of kinetic theory. We refer to the chapter by Y. Elskens, in this book, for further details.

5   Triple collisions and central configurations

So far we have focused on the mechanism of binary collisions. The degree of difficulty in dealing with more complicated models increases immediately as soon as we are concerned with *triple* collisions (see the chapter by J. Waldvogel, in this book). Indeed, one finds an extremely chaotic behaviour, such that a small variation of the initial conditions leads to large effects on the successive dynamics. While regularization always works for binary collisions, triple collisions cannot be regularized, except for a negligible set of masses (see [12], [13]). A remarkable result on N–body collisions is due to K.F. Sundman ([24]), which involves the definition of *total angular momentum*. Let m_j, $j = 1, ..., N$, be the masses of the N bodies and let $x_j \in \mathbf{R}^3$ be their barycentric positions.

Definition: We define the total angular momentum (with respect to the origin) of N bodies as the quantity

$$C = \sum_{j=1}^{N} m_j \, x_j \wedge \dot{x}_j \,,$$

where $x \wedge y$ denotes the cross product between the vectors x and y.

A necessary condition for collisions is stated by the following

Theorem (Sundman): If at a time $t = t_c$ all bodies collide at the origin, then $C = 0$ for all times before the collision time t_c.

A key role in the study of triple collisions is played by the so–called *central configurations*, which we are going to define as follows. Let $U \equiv U(x)$ ($x = (x_1, ..., x_N) \in \mathbf{R}^{3N} \setminus \bigcup_{1 \le j < k \le N} \{x \in \mathbf{R}^{3N} / x_j = x_k\}$) be the force function

$$U(x) \equiv \sum_{1 \le j < k \le N} \frac{m_j m_k}{|x_j - x_k|} \,;$$

then the equations of motion can be written as

$$m_j \ddot{x}_j = \frac{\partial U(x)}{\partial x_j} \qquad \text{for } j = 1, ..., N \,.$$

Definition: A configuration $x \in \mathbf{R}^{3N}$ is called a central configuration, if there exists a constant μ such that

$$\nabla U(x) = \mu \, M x \,,$$

where M is the $3N \times 3N$ diagonal matrix

$$M = \text{diag}(m_1, m_1, m_1, ..., m_N, m_N, m_N) .$$

For $N = 3$ examples of central configurations are given by the collinear (Eulerian) and the triangular (Lagrangian) configurations. The latter case corresponds to the position occupied by the Trojan and Greek asteroids forming an equilateral triangle with Jupiter and the Sun.

Definition: A solution $x = x(t)$ is called *homographic*, if the N bodies form a configuration which remains similar to itself for any time. In formulae: there exist a positive real function $r = r(t)$ and a 3×3 orthogonal matrix $\Omega(t)$ such that

$$x_i(t) = r(t) \, \Omega(t) \, x_i(0) , \qquad i = 1, ..., N .$$

Definition: A homographic solution $x = x(t)$ is called *homothetic,* if the configuration expands or shrinks without rotation, i.e. if $\Omega(t)$ reduces to the identity matrix.

The relation between central configurations and homographic solutions is given by the following statement, which generalizes the result on the Lagrangian solutions for $N = 3$: a necessary and sufficient condition for homographic solutions to occur is that the N bodies form at any instant the same central configuration.

A geometrical description of triple collisions is provided by the following result (see [24]):

Theorem: The solutions tending to a triple collision asymptotically approach a central configuration.

We refer the reader to [2] for further discussion of this subject.

6   Chaotic diffusion: the inclined billiard

The procedure to approach a realistic physical problem is to start from a simple approximation. Along this direction, the investigation of the dynamics of the solar system starts from the simplest case of the two–body approximation, for which the solution is given by the Keplerian laws. The next step requires to add one more planet, leading thus to the three–body problem. Step by step, one adds new degrees of difficulty, namely more bodies of the solar system (planets, satellites, comets, asteroids), the asphericity of such bodies, tides, solar wind and so on.

However, even the simplest nontrivial model may present difficulties in the analytical treatment of the equations of motion. Indeed, we know that the three–body problem, even the restricted, circular, planar model, is known to be non–integrable after H. Poincaré ([18]). However, due to the fact that the mass ratio of any planet to the Sun is rather small (at most of the order of 10^{-3} in the case of Jupiter), the restricted three–body problem is almost integrable, with the perturbing parameter given by the mass ratio of the primaries. We can therefore apply perturbation techniques to get an approximation of the equations of motion. From the numerical point of view, one can reduce the numerical integration of the equations of motion to the computation of the Poincaré map, which drastically reduces the computational effort as well as the difficulties in the interpretation of the numerical inspection of the dynamics.

Such a procedure of reducing a problem to a simple model which retains all the essential features of the original problem is very common in Celestial Mechanics and it was adopted by M. Hénon to describe collisions and chaotic scattering of gravitational systems. His model is known as the *inclined billiard* (see [8], [9], [17] and the chapter by J.–M. Petit, in this book), and it deals with the phenomenon of *diffusion*, occurring whenever two particles interact in a complicated way and separate after some time. Let us remark that the scattering problem is common to many other fields of physics: from molecular physics, concerning the collision between an atom and a molecule, to the scattering of an electron in the Earth's magnetic field, to geophysics as charged particles emitted from the Sun come closer to the Earth.

However, the most striking example is provided by Saturn's rings, which behave like a chaotic dynamics due to the extreme sensitivity to initial conditions. Let us consider the case of two particles of Saturn's rings, say P_1 and P_2, with comparable masses; they describe almost circular orbits. This model is closely related to *Hill's problem*. The interaction of the two minor bodies, and their possible collision, leads to a very complicated dynamical behaviour. Let us denote by h their minimal or *impact* distance. As far as h is large, the trajectories are weakly perturbed, while for h small one observes a horseshoe orbit. Regular and transition regions are the outcome of a numerical investigation. In the transition regime, the trajectory changes drastically even for small deviations of the impact parameter; the experiments show discontinuities in the values of h at which the orbit undergoes drastic changes. However, magnifying a transition region, one observes that it is itself divided into regular and transition regimes, providing a self-similar (or Cantor) structure. Hill's problem presents an intrinsic difficulty, due to the fact that the asymptotic periodic orbits have very large eigenvalues, thus leading to a highly unstable motion. On the basis of this remark, M. Hénon developed a simple model, the inclined billiard with gravity, which has the essential features of Hill's problem. The billiard table is assumed to be an inclined plane, and the particle bounces elastically on two disks. This particle, subject to a constant acceleration, is reduced to a point–mass without rotational effects and friction is neglected. After its release with zero initial velocity, the particle can have different behaviours according to the initial position (see Fig. 4).

If the particle starts exactly on the vertical drawn from A or B, the trajectory is exactly periodic. Let $(x_0, y_0) = (h, 0)$ be the initial conditions, with h denoting the impact distance. For small variations of h, one can have an escape on the left or on the right, or a bouncing between the two disks (followed possibly by an es-

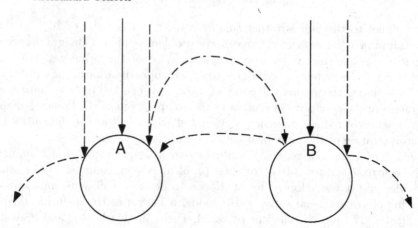

Fig. 4. The inclined billiard.

cape). The limit of large disks is also considered, and the corresponding problem is described by a piecewise linear mapping. The overall picture of the dynamics associated with this simple problem is very similar to that of the particles in Saturn's rings, showing regular and transition regions, discontinuities, self–similar structure. Symbolic dynamics according to the direction of the rebounding has been introduced to describe the model. More precisely, any trajectory can be represented by a sequence of 0's or 1's, where 0 corresponds to a rebounding to the left and 1 to the right. Having fixed an arbitrary sequence of reboundings, one can construct an orbit with the specified properties. Such a *Bernoulli system* is highly chaotic. We refer to the chapter by J.–M. Petit, in this book, for further details.

7   Noncollision singularities

So far, we discussed the problem of singularities in gravitational systems which are due to collisions. However, collisions might not be the only source of singularities. In other words, one might ask whether there exist motions which become unbounded in finite time. This concept might seem counterintuitive and puzzled celestial mechanicians for more than a century. We anticipate that the question has been recently solved by Z. Xia ([27]) for a 5–body problem, but it still remains open in the case of four bodies. The fact that a body can go to infinity in finite time can be understood by means of the following example: suppose that a particle is repeatedly accelerated by a slingshot effect, such that it covers a unit distance in 1 second, the second unit distance in half time, than in 1/4 of a second and so on. The time to reach infinity is given by the geometric series $\sum_{j=0}^{\infty} \frac{1}{2^j}$, which is equal to 2 seconds!

Let us start with the precise definition of noncollision singularities (for a more detailed description of the results, see the second chapter by C. Falcolini, in this book).

Definition: A collision singularity occurs at time $t = t_c$ if the position vector x has a definite limit as t tends to t_c. A noncollision singularity (or *pseudocollision)* occurs at time $t = t_c$ if the position vector x is unbounded as t tends to t_c.

At the beginning of the XX century H. von Zeipel showed that a system can experience a noncollision singularity, only if the motion becomes unbounded in finite time. More precisely, denoting the polar *moment of inertia* (with respect to the origin) of an N–body system as the quantity

$$I(x) \equiv \frac{1}{2} \sum_{j=1}^{N} m_j |x_j|^2 \, ,$$

von Zeipel proved that if

$$\lim_{t \to t_c} I(x(t))$$

is finite for some t_c being a singularity, then the singularity is due to a collision ([31]). The converse is also true, in the sense that a noncollision singularity can occur if one or more bodies escape to infinity in finite time. In this case, the escaping body should acquire infinite kinetic energy, which seems impossible. However, Z. Xia provided an explanation by noting that since there is no lower bound on the potential energy, there is no upper bound on the kinetic energy.

The first step toward understanding noncollision singularities was provided by Paul P. Painlevé (Paris, 1863–1933). Painlevé combined a scientific and political career. He studied at the École Normale Supérieure and became member of the Académie des Sciences in 1900. Ten years later he entered the French parliament, attending especially to military problems. He was appointed minister several times, and he was even a candidate for the presidency of the republic. Due to his scientific and political achievements, he was buried in the Pantheon in Paris. His works on the gravitational singularities gave a strong impulse toward the comprehension of the problem. In particular, he proved that in the case of the three–body problem a singularity is always due to a collision. A century–old conjecture due to him opened the way to the study of noncollision singularities:

Painlevé's conjecture: The N–body problem with $N \geq 4$ admits noncollision singularities.

Triple collisions play an important role in the proof of Painlevé's conjecture. A first remark is due to V. Szebehely ([25]) in the framework of the so–called *Pythagorean problem,* which concerns the problem of finding the motion of three bodies of masses 3, 4, 5 initially at rest at the vertices of a Pythagorean triangle with sides 3, 4, 5 as in Fig. 5.

A numerical integration of the equations of motion starting with zero initial velocities shows that a near–triple collision occurs after some time. Afterwards, a binary system composed of the masses 4 and 5 appears, while the third mass escapes to infinity with high velocity. The triple approach provides a sort of slingshot effect, such that the third particle gets farther from the binary. Such

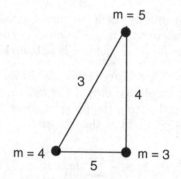

Fig. 5. The Pythagorean problem.

behaviour in the context of the planar three–body problem was also discussed by J. Waldvogel ([29], [30]).

The idea pursued by most people involved in Painlevé's conjecture was to build a system of $N \geq 4$ bodies, where one particle oscillates back and forth between the others, increasing their distances at each oscillation, so that the motion becomes unbounded in finite time. Evidence that collisions may accumulate until a body is ejected to infinity in finite time was given by J. Mather and R. McGehee ([14]). They considered a 4–body problem on a straight line (see Fig. 6); suppose that particles 1 and 2 are close together, well spaced from particle 4, while the third one travels back and forth. At each near–triple collision of bodies 1, 2, 3, the third particle is expelled toward 4. A collision between 3 and 4 rebounds 3 toward 1 and 2. If the next near–triple collision occurs at the right time, particle 3 is again ejected toward 4 with even higher velocity, until the size of the system (i.e., the moment of inertia) increases to infinity in finite time. This is an example of a pseudocollision singularity, except that no collisions among the particles should occur!

Fig. 6. The 4–body example by Mather and McGehee.

A 5–body attempt to settle Painlevé's conjecture was presented by J.L. Gerver in [6]. However, the first complete example of noncollision singularities was successfully provided by Z. Xia ([27]), and it involves at least 5 bodies. His powerful model is the following (see the chapter by C. Falcolini, in this book, for further details). Consider five bodies with masses $m_1 = m_2$, m_3, $m_4 = m_5$. Suppose that bodies 1, 2 and 4, 5 are in parallel planes, where each pair moves on a highly eccentric elliptical orbit. Particle 3 moves along an axis perpendicular to the planes of motion of the other bodies. The third body must arrive at the right time to engage in a near–triple collision with 1, 2, and it must be rebounded toward 4, 5 with sufficiently large velocity. An analogous effect must be achieved by particles 4, 5, rebounding 3 toward 1, 2, until it gains enough kinetic energy to be ejected to infinity in finite time.

A similar example of noncollision singularity involving $3N$ bodies for a large N was presented in [7]. However, there still remains the question of whether Painlevé's conjecture is valid for $N = 4$. At present time no proof is available, and we leave any attempt to the reader...

Acknowledgements

I am deeply indebted to G. Bellettini, C. Falcolini, C. Froeschlé, Ch. Froeschlé, G.F. Gronchi, F. Nicoló, L. Triolo, J. Waldvogel for their helpful comments and suggestions during the accomplishment of this work.

References

1. G.D. Birkhoff: Trans. Am. Math. Soc. **18**, 199 (1917)
2. F.N. Diacu: *Singularities of the N–Body Problem: an Introduction to Celestial Mechanics*. (Les Publications CRM, Montréal 1992)
3. C.C. Dyer, P.S.S. Ip: Astroph. J. **409**, 60 (1993)
4. R. Easton: J. Diff. Eq. **10**, 92 (1971)
5. G.L. Gerstein, D.H. Perkel, J.E. Dayhoff: J. of Neuroscience **5**, n. 4, 881 (1985)
6. J.L. Gerver: J. Diff. Eq. **52**, 76 (1984)
7. J.L. Gerver: J. Diff. Eq. **89**, 1 (1991)
8. M. Hénon: Physica D **33**, 132 (1988)
9. M. Hénon: La Recherche **20**, 490 (1989)
10. P. Kustaanheimo, E. Stiefel: J. Reine Angew. Math **218**, 204 (1965)
11. B.G. Lindsey, R. Shannon, G.L. Gerstein: Brain Research **483**, 373 (1989)
12. R. McGehee: Invent. Math. **27**, 191 (1974)
13. R. McGehee: in *Dynamical Systems Theory and Applications* (J. Moser, ed.) 550 (1975)
14. J. Mather, R. McGehee: in *Lecture Notes in Physics* (J. Moser, ed.), Springer–Verlag, 573 (1975)
15. D. Merrit: Astron. J. **111**, n. 6, 2462 (1996)
16. P. Moore: *The Guinness Book of Astronomy*. (Patrick Moore and Guinness Publishing Ltd, 1988)
17. J.-M. Petit, M. Hénon: Icarus **66**, 536 (1986)
18. H. Poincaré: *Les Méthodes Nouvelles de la Méchanique Céleste*. (Gauthier Villars, Paris 1892)
19. G.L. Schroeder: *Genesis and the Big Bang*. (Bantam books, New York, Toronto, London, Sydney, Auckland 1990)
20. C.L. Siegel, J.K. Moser: *Lectures on Celestial Mechanics*. (Springer–Verlag, Berlin, Heidelberg, New York 1971)
21. E.L. Stiefel, M. Rössler, J. Waldvogel, C.A. Burdet: NASA Contractor Report **CR-769**, Washington, 1967
22. E.L. Stiefel, G. Scheifele: *Linear and Regular Celestial Mechanics*. (Springer–Verlag, Berlin, Heidelberg, New York 1971)
23. E.L. Stiefel, J. Waldvogel: C. R. Acad. Sc. Paris **260**, 805 (1965)
24. K.F. Sundman: Acta Soc. Sci. Fennicae **34**, n. 6 (1907)

25. V. Szebehely: Proc. Nat. Acad. Sci. U.S.A. **58**, 60 (1967)
26. V. Szebehely: *Theory of Orbits*. (Academic Press, New York and London 1967)
27. Z. Xia: Annals of Math. **135**, 411 (1992)
28. J. Waldvogel: Bull. Astron. **3**, II, 2, 295 (1967)
29. J. Waldvogel: Cel. Mech. **11**, 429 (1975)
30. J. Waldvogel: Cel. Mech. **14**, 287 (1976)
31. H. von Zeipel: Arkiv für Mat. Astr. Fys. **4**, n. 32, 1 (1908)

The Levi–Civita, KS and Radial–Inversion Regularizing Transformations

Alessandra Celletti

Dipartimento di Matematica, Università di Roma "Tor Vergata", Via della Ricerca Scientifica - I-00133 Roma (Italy)

Abstract. We review the Levi–Civita, Kustaanheimo–Stiefel and radial–inversion regularizing transformations. The Levi–Civita technique is used to deal with planar motions and its extension to the spatial case is the Kustaanheimo–Stiefel transformation. An alternative procedure is provided by the so–called radial–inversion transformation. In all cases, the basic tool is to perform suitable coordinate and time transformations in the extended phase space. We apply the Levi–Civita, Kustaanheimo–Stiefel and radial–inversion transformations to the two-body problem and to the restricted three-body problem. The Hamiltonian formalism is used, which ensures the canonicity of the transformations.

1 Introduction

Consider two or more massive bodies under the effect of the mutual gravitational attraction. It can happen that these bodies experience a *collision* at a given time. When dealing with the two-body problem, the bodies can collide if they move on a straight line. Due to the fact that the newtonian force depends on the inverse of the square of the distance, a collision implies a singularity in the equations of motion, since during the collision the distance becomes zero. The aim of regularization theory is to transform singular differential equations into regular ones. This method is also useful for the numerical solution of close encounter–like problems, since the regularized equations can be integrated using a larger time step.

In this chapter we review the Levi–Civita, Kustaanheimo–Stiefel (hereafter, KS) and radial–inversion regularization theories, which can be used to investigate two or three-body problems. The Levi–Civita transformation is applied when the bodies are assumed to move on a plane; the spatial case is covered by the Kustaanheimo–Stiefel and radial–inversion transformations. All these theories encompass the case of a collision with one body; in the three-body context, the simultaneous regularization with both primaries is performed in [1], [6], [7], [8], [10].

For completeness we shortly review the basic Hamiltonian equations describing the two and three-body motions. In particular, we shall mainly confine ourselves to the study of the *planar, circular, restricted* three-body problem. For the review of the Levi–Civita and Kustaanheimo–Stiefel regularization, we strictly follow the theory developed in [7], [9], providing as many details as possible, while we refer to [5] for more details on the radial–inversion transformation.

We found extremely useful to adopt the Hamiltonian formalism, which allows to control the conservation of the integrals and to check the canonicity of the transformations.

2   The two- and three-body problem

In this chapter, we shortly review the basic equations describing the two and three-body problems, referring to [2], [9] for an exhaustive description. In particular, we consider a special class of three-body motion, namely the planar, circular, restricted three-body problem.

2.1   The two-body problem

Let $\overline{P}_1$, $\overline{P}_2$ be two massive bodies attracting each other by a newtonian force. According to Kepler's laws, the motion takes place on a plane. In a fixed reference frame with relative cartesian coordinates (q_1, q_2) in the plane of motion, the two-body problem is described in suitable units of measure by the Hamiltonian

$$H(p_1, p_2, q_1, q_2) = \frac{1}{2}(p_1^2 + p_2^2) - \frac{1}{(q_1^2 + q_2^2)^{\frac{1}{2}}} \, ,$$

where $p_j = \dot{q}_j$, $j = 1, 2$. The corresponding Hamilton's equations are:

$$\dot{q}_1 = \frac{\partial H}{\partial p_1} = p_1 \qquad \dot{p}_1 = -\frac{\partial H}{\partial q_1} = -\frac{q_1}{(q_1^2 + q_2^2)^{\frac{3}{2}}}$$

$$\dot{q}_2 = \frac{\partial H}{\partial p_2} = p_2 \qquad \dot{p}_2 = -\frac{\partial H}{\partial q_2} = -\frac{q_2}{(q_1^2 + q_2^2)^{\frac{3}{2}}} \, .$$

2.2   The planar, circular, restricted three-body problem

Consider a body S of infinitesimal mass subject to the gravitational attraction of two bodies $\overline{P}_1$, $\overline{P}_2$ with masses, respectively, μ_1, μ_2. Such model is usually referred to as the *restricted* problem, since the primary bodies are not affected by the gravitational attraction of S. We assume that the motion of all bodies takes place on the same plane and that $\overline{P}_1$, $\overline{P}_2$ move on circular orbits around their common center of mass. We refer to this model as the *planar, circular, restricted three-body problem*. On the plane of motion, let (q_1, q_2) be the cartesian coordinates of S with respect to an inertial reference frame centered at the barycenter of $\overline{P}_1$ and $\overline{P}_2$. We assume that the units of measure are chosen so that

$$\mu_1 + \mu_2 = 1$$

and let the coordinates of the other bodies be $\overline{P}_1(X_1, Y_1)$, $\overline{P}_2(X_2, Y_2)$. The Lagrangian function associated to the motion of S is

$$L(\dot{q}_1, \dot{q}_2, q_1, q_2, t) = \frac{1}{2}(\dot{q}_1^2 + \dot{q}_2^2) + V(q_1, q_2, t) \, ,$$

where

$$V\,(q_1\,,q_2,t) \equiv \frac{\mu_1}{\rho_1} + \frac{\mu_2}{\rho_2}$$

$$\rho_1 \equiv \sqrt{(q_1 - X_1)^2 + (q_2 - Y_1)^2}\,, \qquad \rho_2 \equiv \sqrt{(q_1 - X_2)^2 + (q_2 - Y_2)^2}\,.$$

Notice that in the above expressions X_1, Y_1, X_2, Y_2 are explicit functions of the time. The corresponding Hamiltonian function is

$$H(p_1, p_2, q_1, q_2, t) = \frac{1}{2}(p_1^2 + p_2^2) - V(q_1, q_2, t)\,,$$

where p_1 and p_2 are the kinetic moments conjugated to q_1 and q_2.

Remark: In the spatial case (namely when the motion of the three bodies is not constrained on a plane) one needs to introduce an extra degree of freedom setting

$$H(p_1, p_2, p_3, q_1, q_2, q_3, t) = \frac{1}{2}(p_1^2 + p_2^2 + p_3^3) - V(q_1, q_2, q_3, t)\,,$$

where

$$V(q_1, q_2, q_3, t) = \frac{\mu_1}{\sqrt{(q_1 - X_1)^2 + (q_2 - Y_1)^2 + (q_3 - Z_1)^2}}$$
$$+ \frac{\mu_2}{\sqrt{(q_1 - X_2)^2 + (q_2 - Y_2)^2 + (q_3 - Z_2)^2}}$$

with $S \equiv S(q_1, q_2, q_3)$, $\overline{P}_1 \equiv \overline{P}_1(X_1, Y_1, Z_1)$, $\overline{P}_2 \equiv \overline{P}_2(X_2, Y_2, Z_2)$.

Coming back to the planar case, we derive the equations of motion of S in a rotating or *synodic* reference frame (see Fig. 1) centered at the barycenter of $\overline{P}_1$ and $\overline{P}_2$; we assume that the units of measure are chosen so that the relative angular velocity of $\overline{P}_1$ and $\overline{P}_2$ is unity.

In such reference system, $\overline{P}_1$ and $\overline{P}_2$ are at rest and their coordinates are $\overline{P}_1(\mu_2, 0)$, $\overline{P}_2(-\mu_1, 0)$. We denote by (Q_1, Q_2) the coordinates of S in the synodic frame. In order to derive the transformed Hamiltonian function, it is convenient to introduce the generating function

$$W(p_1, p_2, Q_1, Q_2, t) = p_1 Q_1 \cos t - p_1 Q_2 \sin t + p_2 Q_1 \sin t + p_2 Q_2 \cos t\,,$$

with associated characteristic equations

$$q_1 = \frac{\partial W}{\partial p_1} = Q_1 \cos t - Q_2 \sin t \qquad P_1 = \frac{\partial W}{\partial Q_1} = p_1 \cos t + p_2 \sin t$$

$$q_2 = \frac{\partial W}{\partial p_2} = Q_1 \sin t + Q_2 \cos t \qquad P_2 = \frac{\partial W}{\partial Q_2} = -p_1 \sin t + p_2 \cos t\,,$$

whose invertion yields

$$Q_1 = q_1 \cos t + q_2 \sin t \qquad p_1 = P_1 \cos t - P_2 \sin t$$
$$Q_2 = -q_1 \sin t + q_2 \cos t \qquad p_2 = P_1 \sin t + P_2 \cos t\,.$$

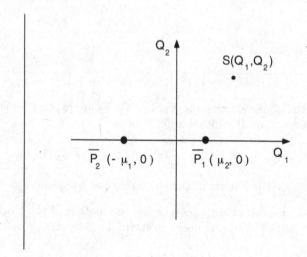

Fig. 1. Position of $\overline{P}_1$, $\overline{P}_2$ and S in the synodic frame.

After some computations, one finds

$$\tilde{H}(P_1, P_2, Q_1, Q_2, t) = H - \frac{\partial W}{\partial t} = \frac{1}{2}(P_1^2 + P_2^2)$$
$$- V(Q_1 \cos t - Q_2 \sin t, Q_1 \sin t + Q_2 \cos t, t) + Q_2 P_1 - Q_1 P_2 .$$

Since in the fixed frame the bodies $\overline{P}_1$ and $\overline{P}_2$ describe circles of radius μ_2 and μ_1, respectively, around the center of mass, their coordinates are

$$X_1 = \mu_2 \cos t \qquad X_2 = -\mu_1 \cos t$$
$$Y_1 = \mu_2 \sin t \qquad Y_2 = -\mu_1 \sin t .$$

Therefore, the perturbing function is

$$V(q_1, q_2, t) \equiv \frac{\mu_1}{\sqrt{(q_1 - \mu_2 \cos t)^2 + (q_2 - \mu_2 \sin t)^2}} + \frac{\mu_2}{\sqrt{(q_1 + \mu_1 \cos t)^2 + (q_2 + \mu_1 \sin t)^2}}$$

and in the new set of variables it reduces to

$$\tilde{V}(Q_1, Q_2) = \frac{\mu_1}{\sqrt{(Q_1 - \mu_2)^2 + Q_2^2}} + \frac{\mu_2}{\sqrt{(Q_1 + \mu_1)^2 + Q_2^2}} .$$

Finally, in the synodic reference frame the Hamiltonian takes the form

$$\tilde{H}(P_1, P_2, Q_1, Q_2) = \frac{1}{2}(P_1^2 + P_2^2) + Q_2 P_1 - Q_1 P_2 - \tilde{V}(Q_1, Q_2) . \qquad (1)$$

Hamilton's equations associated to (1) are

$$\dot{Q}_1 = P_1 + Q_2 \qquad \dot{P}_1 = P_2 + \tilde{V}_{Q_1}$$
$$\dot{Q}_2 = P_2 - Q_1 \qquad \dot{P}_2 = -P_1 + \tilde{V}_{Q_2} ,$$

from which it follows that

$$\ddot{Q}_1 - 2\dot{Q}_2 = \overline{\Omega}_{Q_1}$$
$$\ddot{Q}_2 + 2\dot{Q}_1 = \overline{\Omega}_{Q_2} \, , \tag{2}$$

where

$$\overline{\Omega} \equiv \frac{1}{2}(Q_1^2 + Q_2^2) + \tilde{V}(Q_1, Q_2) \, . \tag{3}$$

Moreover, let $\Omega \equiv \overline{\Omega} + \frac{1}{2}\mu_1\mu_2$. We recall the *Jacobi integral* as follows: multiplying the first equation in (2) by $\dot{Q}_1$ and the second by $\dot{Q}_2$ and adding the two equations, one obtains:

$$\dot{Q}_1^2 + \dot{Q}_2^2 = 2\overline{\Omega} - C' = 2\Omega - C \, .$$

Let the *Jacobi constant* be defined as

$$C \equiv 2\Omega - (\dot{Q}_1^2 + \dot{Q}_2^2) \, .$$

Remark: Since $\dot{Q}_1 = P_1 + Q_2$ and $\dot{Q}_2 = P_2 - Q_1$, one has $P_1 = \dot{Q}_1 - Q_2$, $P_2 = \dot{Q}_2 + Q_1$; therefore, the Hamiltonian in terms of $Q_1, Q_2, \dot{Q}_1, \dot{Q}_2$ becomes

$$\tilde{H} = \frac{1}{2}(\dot{Q}_1^2 + \dot{Q}_2^2) - \frac{1}{2}(Q_1^2 + Q_2^2) - V(Q_1, Q_2) = \frac{1}{2}(\dot{Q}_1^2 + \dot{Q}_2^2) - \overline{\Omega} \, .$$

Using the Jacobi integral one obtains

$$\tilde{H} = \frac{\mu_1\mu_2 - C}{2} \, .$$

3   The Levi–Civita regularization

Let us start with the simplest cases of two or three bodies moving on a *plane*. In such case, the regularizing transformation is provided by the Levi–Civita method. We first show the effect of such technique on the two-body system. Once this problem is solved, its generalization to the three-body system becomes quite straightforward. The basic steps of the Levi–Civita regularization are provided by a coordinate change of variables (known as the *Levi–Civita transformation*), the introduction of the *extended phase space* to get rid of the time dependence and the introduction of a fictitious time.

3.1   The two-body problem

We recall that the Hamiltonian of the two-body problem is given by

$$H(p_1, p_2, q_1, q_2) = \frac{1}{2}(p_1^2 + p_2^2) - \frac{1}{(q_1^2 + q_2^2)^{\frac{1}{2}}} \, .$$

In order to regularize the associated equations of motion, let us introduce a canonical transformation of coordinates with generating function which we assume to be linear in the p-variables:

$$W(p_1, p_2, Q_1, Q_2) = p_1 f(Q_1, Q_2) + p_2 g(Q_1, Q_2) .$$

The functions f and g corresponding to the *Levi–Civita transformation* are given by

$$f(Q_1, Q_2) \equiv Q_1^2 - Q_2^2 , \qquad\qquad g(Q_1, Q_2) \equiv 2Q_1 Q_2 ;$$

denoting by $i = \sqrt{-1}$ the imaginary unit, one has

$$f + ig \equiv (Q_1 + iQ_2)^2 = Q_1^2 - Q_2^2 + i \cdot 2Q_1 Q_2 .$$

The characteristic equations associated to the generating function W are

$$q_1 = \frac{\partial W}{\partial p_1} = f(Q_1, Q_2) = Q_1^2 - Q_2^2$$

$$q_2 = \frac{\partial W}{\partial p_2} = g(Q_1, Q_2) = 2Q_1 Q_2$$

$$P_1 = \frac{\partial W}{\partial Q_1} = p_1 \frac{\partial f}{\partial Q_1} + p_2 \frac{\partial g}{\partial Q_1} = 2p_1 Q_1 + 2p_2 Q_2$$

$$P_2 = \frac{\partial W}{\partial Q_2} = p_1 \frac{\partial f}{\partial Q_2} + p_2 \frac{\partial g}{\partial Q_2} = -2p_1 Q_2 + 2p_2 Q_1 .$$

Let the *physical plane* be described by the coordinates (q_1, q_2) and the *parametric plane* by (Q_1, Q_2), which are related through $q_1 + iq_2 = f + ig = (Q_1 + iQ_2)^2$. We remark that applying the Levi–Civita transformation the angles at the orgin are doubled; therefore, a particle which makes a revolution around the center of mass is transformed to a point of the parametric plane which has made half revolution.

The last two equations of the above transformation can be inverted as

$$P = 2A_0^+ \, p \qquad \text{with} \qquad A_0 = \begin{pmatrix} Q_1 & -Q_2 \\ Q_2 & Q_1 \end{pmatrix}$$

(the superscript + denotes matrix transposition). Let $D = D(Q_1, Q_2) \equiv 4 \det A_0 = 4(Q_1^2 + Q_2^2) > 0$; using the above relations, one obtains

$$P_1^2 + P_2^2 = D(p_1^2 + p_2^2) .$$

Therefore, the new Hamiltonian becomes

$$\tilde{H}(P_1, P_2, Q_1, Q_2) = \frac{1}{2D}(P_1^2 + P_2^2) - \frac{1}{(f(Q_1, Q_2)^2 + g(Q_1, Q_2)^2)^{\frac{1}{2}}} ,$$

with Hamilton's equations:

$$\dot{Q}_1 = \frac{P_1}{D}$$

$$\dot{Q}_2 = \frac{P_2}{D}$$

$$\dot{P}_1 = \frac{1}{2D}(P_1^2 + P_2^2)\frac{\partial D}{\partial Q_1} - \frac{1}{2}\frac{1}{(f^2+g^2)^{\frac{3}{2}}}\frac{\partial(f^2+g^2)}{\partial Q_1}$$

$$\dot{P}_2 = \frac{1}{2D}(P_1^2 + P_2^2)\frac{\partial D}{\partial Q_2} - \frac{1}{2}\frac{1}{(f^2+g^2)^{\frac{3}{2}}}\frac{\partial(f^2+g^2)}{\partial Q_2} .$$

Let us rewrite the Hamiltonian function in the *extended phase space* ([7]), by adding a new pair of conjugated variables (T,t):

$$\Gamma(P_1, P_2, T, Q_1, Q_2, t) = \frac{1}{2D}(P_1^2 + P_2^2) + T - \frac{1}{(f(Q_1,Q_2)^2 + g(Q_1,Q_2)^2)^{\frac{1}{2}}} .$$

Notice that $\dot{t} = \frac{\partial \Gamma}{\partial T} = 1$ and $\dot{T} = -\frac{\partial \Gamma}{\partial t} = 0$, so that $T = $ const. $= -\tilde{H}$; in particular along a solution one obtains $T(t) = -\tilde{H}$.

Remark: The extended phase space is introduced so to obtain a transformation involving also the time. In general, if $\tilde{H} = \tilde{H}(P,Q,t)$ depends explicitly on the time, one can introduce a time–independent Hamiltonian $\Gamma = \Gamma(P,Q,T,t) \equiv \tilde{H}(P,Q,t) + T$, with T being conjugated to t. The Hamiltonian Γ is identically zero along any solution provided that $T(0) = -\tilde{H}(P(0), Q(0))$.

We next introduce a *fictitious time* or *regularized time* s defined as

$$dt = D(Q_1, Q_2)ds \qquad \text{or} \qquad \frac{d}{dt} = \frac{1}{D}\frac{d}{ds} .$$

Let us derive the transformed Hamiltonian as follows. From $\dot{Q} = \frac{dQ}{dt} = \frac{dQ}{ds}\frac{ds}{dt} = \frac{1}{D}\frac{dQ}{ds} = \frac{\partial \Gamma}{\partial P}$, one has

$$\frac{dQ}{ds} = \frac{\partial \Gamma^*}{\partial P} ,$$

with $\Gamma^* \equiv D\Gamma$. As for P, one has $\dot{P} = \frac{dP}{dt} = \frac{dP}{ds}\frac{ds}{dt} = \frac{1}{D}\frac{dP}{ds} = -\frac{\partial \Gamma}{\partial Q}$, so that

$$\frac{dP}{ds} = -\frac{\partial \Gamma^*}{\partial Q}$$

with $\Gamma^* \equiv D\Gamma$, since

$$\frac{\partial \Gamma^*}{\partial Q} = \frac{\partial D}{\partial Q}\Gamma + D\frac{\partial \Gamma}{\partial Q} = D\frac{\partial \Gamma}{\partial Q} ,$$

being $\Gamma = 0$ along a solution.

Therefore, the new Hamiltonian Γ^* becomes

$$\Gamma^* \equiv D\Gamma = DT + \frac{1}{2}(P_1^2 + P_2^2) - \frac{D}{(f^2+g^2)^{\frac{1}{2}}}$$

with corresponding Hamilton's equations ($j = 1, 2$):

$$\frac{dQ_j}{ds} = P_j$$

$$\frac{dP_j}{ds} = -\frac{\partial}{\partial Q_j}[DT - \frac{D}{(f^2 + g^2)^{\frac{1}{2}}}]$$

$$\frac{dt}{ds} = D$$

$$\frac{dT}{ds} = 0 . \tag{4}$$

Notice that the singularity of the problem is associated to the term $\frac{D}{(f^2+g^2)^{\frac{1}{2}}}$, which is transformed to

$$\frac{D}{(f^2 + g^2)^{\frac{1}{2}}} = \frac{D}{r} = \frac{4(Q_1^2 + Q_2^2)}{(Q_1^4 + Q_2^4 - 2Q_1^2 Q_2^2 + 4Q_1^2 Q_2^2)^{\frac{1}{2}}} = 4 .$$

The desired regularization is achieved, since the singularity has been removed from the equations of motion with respect to the fictitious time.

In the case of the two-body problem the regularized equations (4) can be easily solved as follows. Denoting by a prime the derivative with respect to s, we rewrite the first two equations in (4) as

$$Q'_j = P_j$$

$$P'_j = -T\frac{\partial D}{\partial Q_j} = -8TQ_j = 8\tilde{H}Q_j \qquad (j = 1, 2) ,$$

being $T = -\tilde{H}$. From the above equations one gets the second order differential equation

$$Q''_j = 8\tilde{H}Q_j \qquad (j = 1, 2) .$$

If $\tilde{H} < 0$ (corresponding to an elliptic orbit), one obtains the equation of an harmonic oscillator. Let $Q_j = Q_j(s)$ be its solution; by expressing s in terms of t through $dt = 4(Q_1^2 + Q_2^2)ds$, one gets $Q_j = Q_j(t)$. Finally, the solution in the original set of coordinates (q_1, q_2) is provided by the expressions:

$$q_1 = q_1(t) = Q_1(t)^2 - Q_2(t)^2 , \qquad q_2 = q_2(t) = 2Q_1(t)Q_2(t) .$$

3.2   The planar, circular, restricted three-body problem

We have shown in 2.2 that the Hamiltonian of the planar, circular, restricted three-body problem is given in a synodic reference frame, by

$$H(p_1, p_2, q_1, q_2) = \frac{1}{2}(p_1^2 + p_2^2) + q_2 p_1 - q_1 p_2 - \tilde{V}(q_1, q_2) ,$$

where $\tilde{V}(q_1, q_2) = \frac{\mu_1}{r_1} + \frac{\mu_2}{r_2}$ with

$$r_1 = [(q_1 - \mu_2)^2 + q_2^2]^{\frac{1}{2}} , \qquad r_2 = [(q_1 + \mu_1)^2 + q_2^2]^{\frac{1}{2}}$$

(notice that we have changed the notation with respect to (1) using small letters instead of capital letters). Consider a canonical transformation

$$(p_1, p_2, q_1, q_2) \rightarrow (P_1, P_2, Q_1, Q_2)$$

defined by a generating function of the form

$$W(p_1, p_2, Q_1, Q_2) = p_1 f(Q_1, Q_2) + p_2 g(Q_1, Q_2) \ .$$

In order to regularize a collisional motion of the small particle S with $\overline{P}_1$, one defines the functions f and g as

$$f(Q_1, Q_2) = Q_1^2 - Q_2^2 + \mu_2 \ , \qquad g(Q_1, Q_2) = 2Q_1 Q_2 \ .$$

In case of collisions with $\overline{P}_2$ replace f by $f(Q_1, Q_2) = Q_1^2 - Q_2^2 - \mu_1$. The characteristic equations are:

$$q_1 = \frac{\partial W}{\partial p_1} = f(Q_1, Q_2)$$

$$q_2 = \frac{\partial W}{\partial p_2} = g(Q_1, Q_2)$$

$$P_1 = \frac{\partial W}{\partial Q_1} = p_1 \frac{\partial f}{\partial Q_1} + p_2 \frac{\partial g}{\partial Q_1}$$

$$P_2 = \frac{\partial W}{\partial Q_2} = p_1 \frac{\partial f}{\partial Q_2} + p_2 \frac{\partial g}{\partial Q_2} \ .$$

Again, the term $p_1^2 + p_2^2$ is transformed into $\frac{1}{D}(P_1^2 + P_2^2)$, while the term $q_2 p_1 - p_2 q_1$ becomes

$$q_2 p_1 - p_2 q_1 = \frac{1}{2D} \ [P_1 \frac{\partial}{\partial Q_2}(f^2 + g^2) - P_2 \frac{\partial}{\partial Q_1}(f^2 + g^2)] \ .$$

Therefore, the new Hamiltonian becomes

$$\tilde{H}(P_1, P_2, Q_1, Q_2) = \frac{1}{2D}[P_1^2 + P_2^2 + P_1 \frac{\partial}{\partial Q_2}(f^2 + g^2) - P_2 \frac{\partial}{\partial Q_1}(f^2 + g^2)] - \hat{V}(Q_1, Q_2) \ ,$$

where $\hat{V}$ is $\tilde{V}$ with $f(Q_1, Q_2)$ in place of q_1 and $g(Q_1, Q_2)$ in place of q_2. The corresponding equations of motion are

$$\dot{Q}_1 = \frac{1}{2D}[2P_1 + \frac{\partial}{\partial Q_2}(f^2 + g^2)]$$

$$\dot{Q}_2 = \frac{1}{2D}[2P_2 - \frac{\partial}{\partial Q_1}(f^2 + g^2)]$$

$$\dot{P}_1 = -\frac{\partial \tilde{H}}{\partial Q_1}$$

$$\dot{P}_2 = -\frac{\partial \tilde{H}}{\partial Q_2} \ .$$

The Hamiltonian in the extended phase space is

$$\Gamma = T + \frac{1}{2D}[P_1^2 + P_2^2 + P_1\frac{\partial}{\partial Q_2}(f^2 + g^2) - P_2\frac{\partial}{\partial Q_1}(f^2 + g^2)] - \hat{V}(Q_1, Q_2) \ .$$

As in the two-body problem, we introduce the fictitious time as

$$dt = D \, ds \ ,$$

which leads to the Hamiltonian

$$\Gamma^* = D\Gamma = DT + \frac{1}{2}[P_1^2 + P_2^2 + P_1\frac{\partial}{\partial Q_2}(f^2 + g^2) - P_2\frac{\partial}{\partial Q_1}(f^2 + g^2)] - D\hat{V}(Q_1, Q_2) \ .$$

Let $\Phi(Q_1, Q_2) \equiv f(Q_1, Q_2) + ig(Q_1, Q_2)$ (with $|\Phi|^2 = f^2 + g^2$); then Hamilton's equations with respect to the fictitious time are

$$Q_1' = P_1 + \frac{1}{2}\frac{\partial}{\partial Q_2}|\Phi|^2$$

$$Q_2' = P_2 - \frac{1}{2}\frac{\partial}{\partial Q_1}|\Phi|^2$$

$$t' = D$$

$$P_1' = -T\frac{\partial D}{\partial Q_1} - \frac{1}{2}[P_1\frac{\partial^2|\Phi|^2}{\partial Q_1\partial Q_2} - P_2\frac{\partial^2|\Phi|^2}{\partial Q_1^2}] + \frac{\partial}{\partial Q_1}(D\hat{V})$$

$$P_2' = -T\frac{\partial D}{\partial Q_2} - \frac{1}{2}[P_1\frac{\partial^2|\Phi|^2}{\partial Q_2^2} - P_2\frac{\partial^2|\Phi|^2}{\partial Q_2\partial Q_1}] + \frac{\partial}{\partial Q_2}(D\hat{V})$$

$$T' = 0 \ .$$

The singularities appear in the term $\frac{\partial}{\partial Q_j}(D\hat{V})$, analogously to the term $D(f^2 + g^2)^{-\frac{1}{2}}$ appearing in the two-body problem. From eq. (3), one has $\frac{1}{2}(f^2 + g^2) + \hat{V} = \Omega - \frac{1}{2}\mu_1\mu_2$ with $q_1 = f$, $q_2 = g$ and $\Omega = \frac{1}{2}\mu_1\mu_2 + \frac{1}{2}(Q_1^2 + Q_2^2) + \hat{V}$. Since $|\Phi|^2 = f^2 + g^2$, one obtains

$$\frac{1}{2}|\Phi|^2 + \hat{V} = \Omega - \frac{1}{2}\mu_1\mu_2 \ .$$

From the relation between $\tilde{H}$ and the Jacobi integral it follows that

$$\tilde{H} = -T = \frac{\mu_1\mu_2 - C}{2} \ ,$$

namely

$$\frac{1}{2}|\Phi|^2 - T + \hat{V} = \Omega - \frac{C}{2} \ .$$

Therefore $D\hat{V} = D(\Omega - \frac{C}{2}) - \frac{1}{2}D|\Phi|^2 + DT$, showing that the critical term is $D(\Omega - \frac{C}{2})$; we now prove that such term does not contain singularities. To this end, let us denote by $z = q_1 + iq_2$ and $w = Q_1 + iQ_2$ the complex coordinates in the physical and parametric plane, respectively. In the physical plane, the primaries are located at $z_1 = \mu_2$ and $z_2 = -\mu_1$; the transformation $z = \mu_2 + w^2$ regularizes the singularity at $\overline{P}_1$, while the transformation $z = -\mu_1 + w^2$ regularizes the singularity at $\overline{P}_2$. The functions $z = \mu_2 + w^2$ and $z = -\mu_1 + w^2$ are known as the "Levi–Civita transformations". Notice that these transformations are said to be *local*, since only one of the two singularities is eliminated. *Global* transformations (regularizing both collisions simultaneously) were developed in [1], [8], [10], [6], [7].

Let us consider the function

$$z \equiv \tilde{f}(w) = w^2 + \mu_2 \,,$$

which transforms the point $\overline{P}_1(\mu_2, 0)$ of the physical plane into the origin of the w–plane, while $\overline{P}_2$ has coordinates $w_{1,2} = \pm i$ (since $w^2 = -\mu_1 - \mu_2 = -1$). The transformation of $U \equiv \Omega - \frac{C}{2}$ in terms of the new complex variable w requires the expressions of r_1 and r_2 in terms of w. Since $r_1 = |z - \mu_2|$ and $r_2 = |z + \mu_1|$, one has $r_1 = |w|^2$, $r_2 = |1 + w^2|$; from

$$\mu_1 r_1^2 + \mu_2 r_2^2 = \mu_1 (z - \mu_2)^2 + \mu_2 (z + \mu_1)^2 = z^2 + \mu_1 \mu_2 \,,$$

it follows that

$$\begin{aligned}
U = \Omega - \frac{C}{2} &= \frac{1}{2}\mu_1\mu_2 + \frac{1}{2}(q_1^2 + q_2^2) + V - \frac{C}{2} \\
&= \frac{1}{2}(\mu_1 r_1^2 + \mu_2 r_2^2) + \frac{\mu_1}{r_1} + \frac{\mu_2}{r_2} - \frac{C}{2} \\
&= \frac{1}{2}\Big[\mu_1 |w|^4 + \mu_2 |1 + w^2|^2\Big] + \frac{\mu_1}{|w|^2} + \frac{\mu_2}{|1 + w^2|} - \frac{C}{2} \,.
\end{aligned}$$

Since $D = 4(Q_1^2 + Q_2^2) = 4|w|^2$, the term $DU = D(\Omega - \frac{C}{2})$ does not contain singularities at $\overline{P}_1$.

Let us conclude this paragraph by looking at the behaviour of the velocities with respect to the fictitious time. The Jacobi integral in the physical space is $|\dot{z}|^2 = 2U$, while in the parametric space it becomes

$$|w'|^2 = 8|w|^2 U \,.$$

Therefore, we have:

$$|w'|^2 = 8\mu_1 + |w|^2 \left[\frac{8\mu_2}{|1 + w|^2} + 4\mu_1 |w|^4 + 4\mu_2 |1 + w^2|^2 - 4C\right] \,.$$

In $\overline{P}_1$ one has $r_1 = 0$, namely $w = 0$, which leads to a finite velocity since $|w'|^2 = 8\mu_1$. In $\overline{P}_2$ one has $r_2 = 0$, namely $w = \pm i$, which implies an infinite velocity: $|w'|^2 = \infty$.

4 The Kustaanheimo–Stiefel regularization

A different approach must be used when the three bodies are allowed to move in the space, rather than in the plane like in the Levi–Civita regularization, whose technique cannot be extended in a straightforward way. More precisely, one cannot transform the 3–dimensional physical space into a 3–dimensional parametric space. As shown in [4] one needs to introduce an extra variable and the transformation will be carried out in a 4–dimensional space. The other main ingredients of the Kustaanheimo–Stiefel (hereafter, KS) regularization are similar to the Levi–Civita method, namely a transformation on coordinates and the introduction of a fictitious time. Following [7], we show in a separate section that the KS transformation is canonical.

4.1 The Kustaanheimo–Stiefel transformation

Let us consider the motion in the *space* of three bodies, S, $\overline{P}_1$, $\overline{P}_2$, the latter two having masses, repectively, μ_1 and μ_2, while S has infinitesimal mass ("restricted" problem). The primaries move in a plane on circular orbits around their common center of mass. In the synodic reference frame their coordinates are $\overline{P}_1(\mu_2, 0, 0)$, $\overline{P}_2(-\mu_1, 0, 0)$. We assume that the plane of motion of the primaries rotates with unit angular velocity about the vertical axis. The Hamiltonian function governing the motion is (see §2.2)

$$H(p_1, p_2, p_3, q_1, q_2, q_3) = \frac{1}{2}(p_1^2 + p_2^2 + p_3^2) + q_2 p_1 - q_1 p_2 - \tilde{V}(q_1, q_2, q_3) \ .$$

Notice that, with abuse of notation, we have written the above Hamiltonian using small letters instead of capital letters as in §2.2. The equations of motion of S under the gravitational influence of $\overline{P}_1$ and $\overline{P}_2$ are

$$\ddot{q}_1 - 2\dot{q}_2 = \Omega_{q_1}$$
$$\ddot{q}_2 + 2\dot{q}_1 = \Omega_{q_2}$$
$$\ddot{q}_3 \qquad\quad = \Omega_{q_3} \ ,$$

where $\Omega = \frac{1}{2}(q_1^2 + q_2^2) + \frac{\mu_1}{r_1} + \frac{\mu_2}{r_2} + \frac{1}{2}\mu_1\mu_2$, with $r_1^2 \equiv (q_1 - \mu_2)^2 + q_2^2 + q_3^2$, $r_2^2 \equiv (q_1 + \mu_1)^2 + q_2^2 + q_3^2$. We first perform a *time* transformation and next a *coordinate* transformation. As in the Levi–Civita regularization we define the fictitious time s by

$$dt = D ds \ ,$$

namely $\frac{d}{dt} = \frac{1}{D}\frac{d}{ds}$. We remark that the second derivatives with respect to t and s are related by

$$\frac{d^2}{dt^2} = \frac{d}{dt}\Big(\frac{1}{D}\frac{d}{ds}\Big) = \frac{1}{D}\frac{d}{ds}\Big(\frac{1}{D}\frac{d}{ds}\Big) = \frac{1}{D^2}\frac{d^2}{ds^2} - \frac{1}{D^3}\frac{dD}{ds}\frac{d}{ds} \ .$$

Therefore the equations of motion with respect to the new time s are

$$\frac{1}{D^2}q_1'' - \frac{1}{D^3}D'q_1' - \frac{2}{D}q_2' = \Omega_{q_1}$$

$$\frac{1}{D^2}q_2'' - \frac{1}{D^3}D'q_2' + \frac{2}{D}q_1' = \Omega_{q_2}$$

$$\frac{1}{D^2}q_3'' - \frac{1}{D^3}D'q_3' = \Omega_{q_3} ,$$

namely

$$Dq_1'' - D'q_1' - 2D^2 q_2' = D^3 \Omega_{q_1}$$
$$Dq_2'' - D'q_2' + 2D^2 q_1' = D^3 \Omega_{q_2}$$
$$Dq_3'' - D'q_3' = D^3 \Omega_{q_3} . \qquad (5)$$

Notice that the singular terms are $D^3 \Omega_{q_1}$, $D^3 \Omega_{q_2}$, $D^3 \Omega_{q_3}$, with Ω_{q_1}, Ω_{q_2}, $\Omega_{q_3} \sim O(\frac{1}{r_1^3})$.

Remark: Let (q_1, q_2) be the physical plane and (u_1, u_2) be the parametric plane. The Levi–Civita transformation can be written as

$$\begin{pmatrix} q_1 \\ q_2 \end{pmatrix} = \begin{pmatrix} u_1 & -u_2 \\ u_2 & u_1 \end{pmatrix} \begin{pmatrix} u_1 \\ u_2 \end{pmatrix} = \begin{pmatrix} u_1^2 - u_2^2 \\ 2u_1 u_2 \end{pmatrix} ,$$

where every element of the matrix $A_0(u) \equiv \begin{pmatrix} u_1 & -u_2 \\ u_2 & u_1 \end{pmatrix}$ is linear in u_1, u_2 and $A_0(u)$ is orthogonal.

The first step in KS–theory is to investigate whether there exists a generalization $A(u)$ of the matrix $A_0(u)$ in $\mathbf{R}^n$, having the following properties:
$i)$ the elements of $A(u)$ are linear homogeneous functions of the u_i;
$ii)$ the matrix is orthogonal, namely
  $a)$ the scalar product of different rows vanishes;
  $b)$ each row has norm $u_1^2 + ... + u_n^2$.
A result by A. Hurwitz ([3]) states that such matrix can only be produced if $n = 1, 2, 4$ or 8, but not $n = 3$. For this reason, we need to map the 3–dimensional physical space into a 4–dimensional parametric space, defining

$$A(u) = \begin{pmatrix} u_1 & -u_2 & -u_3 & u_4 \\ u_2 & u_1 & -u_4 & -u_3 \\ u_3 & u_4 & u_1 & u_2 \\ u_4 & -u_3 & u_2 & -u_1 \end{pmatrix} .$$

The extension of the 3–dimensional physical space to a 4–dimensional space is carried out setting the fourth component equal to zero, i.e. $(q_1, q_2, q_3, 0)$.

We introduce the KS regularization for a collision with the primary $\overline{P}_1$ as follows (the collision with $\overline{P}_2$ can be treated in a similar way). Let

$$\begin{pmatrix} q_1 \\ q_2 \\ q_3 \\ 0 \end{pmatrix} = A(u) \begin{pmatrix} u_1 \\ u_2 \\ u_3 \\ u_4 \end{pmatrix} + \begin{pmatrix} \mu_2 \\ 0 \\ 0 \\ 0 \end{pmatrix}$$

$$= \begin{pmatrix} u_1 & -u_2 & -u_3 & u_4 \\ u_2 & u_1 & -u_4 & -u_3 \\ u_3 & u_4 & u_1 & u_2 \\ u_4 & -u_3 & u_2 & -u_1 \end{pmatrix} \begin{pmatrix} u_1 \\ u_2 \\ u_3 \\ u_4 \end{pmatrix} + \begin{pmatrix} \mu_2 \\ 0 \\ 0 \\ 0 \end{pmatrix} , \tag{6}$$

namely

$$q_1 = u_1^2 - u_2^2 - u_3^2 + u_4^2 + \mu_2$$
$$q_2 = 2u_1u_2 - 2u_3u_4$$
$$q_3 = 2u_1u_3 + 2u_2u_4 .$$

Notice that the fourth equation is trivially zero.

Remarks:
1) Notice that for $u_3 = u_4 = 0$ the KS–transformation reduces to the Levi–Civita transformation.
2) The norms of each row of the matrix A are equal to the square of the norm of the vector u: $u_1^2 + u_2^2 + u_3^2 + u_4^2$.
3) If one wants to regularize $\overline{P}_2$ instead of $\overline{P}_1$, it suffices to substitute the constant vector $(\mu_2, 0, 0, 0)$ with $(-\mu_1, 0, 0, 0)$.
4) The matrix A is orthogonal: $A^+(u)A(u) = (u, u) \cdot Id$. From this relation, setting $Q \equiv (q_1 - \mu_2, q_2, q_3, 0)$ it follows that

$$r_1^2 = (Q, Q) = Q^+ Q = u^+ A^+(u)A(u)u = (\acute{u}, u)^2 ,$$

namely $r_1 = (u, u) = |u|^2 = u_1^2 + u_2^2 + u_3^2 + u_4^2$.
5) It can be explicitly verified that $A(u)' = A(u')$. As a consequence, $Q' = A(u')u + A(u)u' = 2A(u)u'$. In fact,

$$q_1' = 2u_1u_1' - 2u_2u_2' - 2u_3u_3' + 2u_4u_4'$$
$$q_2' = 2u_2u_1' + 2u_1u_2' - 2u_4u_3' - 2u_3u_4'$$
$$q_3' = 2u_3u_1' + 2u_1u_3' + 2u_4u_2' + 2u_2u_4' , \tag{7}$$

namely

$$\begin{pmatrix} q_1' \\ q_2' \\ q_3' \\ 0 \end{pmatrix} = 2A(u)u' = 2 \begin{pmatrix} u_1u_1' - u_2u_2' - u_3u_3' + u_4u_4' \\ u_2u_1' + u_1u_2' - u_4u_3' - u_3u_4' \\ u_3u_1' + u_1u_3' + u_4u_2' + u_2u_4' \\ u_4u_1' - u_3u_2' + u_2u_3' - u_1u_4' \end{pmatrix} ,$$

where the last equation is named *bilinear relation*:

$$u_4u_1' - u_3u_2' + u_2u_3' - u_1u_4' = 0 .$$

6) Concerning the second derivative, one has

$$Q'' = 2A(u)u'' + 2A(u')u' ,$$

namely

$$q_1'' = 2(u_1 u_1'' - u_2 u_2'' - u_3 u_3'' + u_4 u_4'') + 2(u_1'^2 - u_2'^2 - u_3'^3 + u_4'^3)$$
$$q_2'' = 2(u_2 u_1'' + u_1 u_2'' - u_4 u_3'' - u_3 u_4'') + 4(u_1' u_2' - u_3' u_4')$$
$$q_3'' = 2(u_3 u_1'' + u_1 u_3'' + u_2'' u_4 + u_2 u_4'') + 4(u_1' u_3' + u_2' u_4')$$
$$0 = 2(u_4 u_1'' - u_3 u_2'' + u_2 u_3'' - u_1 u_4'') \tag{8}$$

(notice that the last equation follows from the bilinear relation).

In order to obtain the KS regularization we remark that the scale factor D is given by

$$D \equiv 4r_1 = 4(u, u) = 4(u_1^2 + u_2^2 + u_3^2 + u_4^2)$$

and that $D' = 4r_1' = 8(u_1 u_1' + u_2 u_2' + u_3 u_3' + u_4 u_4')$. Therefore, the equations of motion are given by (5) where $q_1, q_2, q_3, q_1', q_2', q_3', q_1'', q_2'', q_3''$ are expressed in terms of u, u', u'' through (6), (7), (8). The singular part of the equations (5) is given by $D^3 \Omega_{q_1}$ (or $D^3 \Omega_{q_2}, D^3 \Omega_{q_3}$). Since $\Omega_{q_1} \propto \frac{1}{r_1^3}$ and $D \propto r_1$, it follows that $D^3 \Omega_{q_1} = O(1)$ and the regularization of the singularity in $\overline{P}_1$ is thus obtained. This concludes the KS regularization; we devote next section to the proof of its canonical character ([7]).

4.2 Canonicity of the KS-transformation

We have seen that in the planar case the KS–transformation reduces to the Levi–Civita transformation:

$$\begin{pmatrix} q_1 \\ q_2 \end{pmatrix} = \begin{pmatrix} u_1 & -u_2 \\ u_2 & u_1 \end{pmatrix} \begin{pmatrix} u_1 \\ u_2 \end{pmatrix} = \begin{pmatrix} u_1^2 - u_2^2 \\ 2u_1 u_2 \end{pmatrix} \equiv A_0(u) u \ ,$$

with $A_0(u) = \begin{pmatrix} u_1 & -u_2 \\ u_2 & u_1 \end{pmatrix}$. If $U = (U_1, U_2)$ is conjugated to $u = (u_1, u_2)$, one obtains

$$p = \begin{pmatrix} p_1 \\ p_2 \end{pmatrix} = \frac{1}{2}(A_0^+)^{-1} U = \frac{1}{2(u_1^2 + u_2^2)} \begin{pmatrix} u_1 & -u_2 \\ u_2 & u_1 \end{pmatrix} \begin{pmatrix} U_1 \\ U_2 \end{pmatrix}$$

$$= \frac{1}{2(u_1^2 + u_2^2)} A_0(u) \begin{pmatrix} U_1 \\ U_2 \end{pmatrix} \ . \tag{9}$$

In the spatial case, we define a 4×3 matrix $\Lambda(u)$ as

$$\Lambda(u) \equiv \begin{pmatrix} u_1 & -u_2 & -u_3 & u_4 \\ u_2 & u_1 & -u_4 & -u_3 \\ u_3 & u_4 & u_1 & u_2 \end{pmatrix} \ ,$$

which is obtained from $A(u)$ suppressing the last row. Let

$$\begin{pmatrix} q_1 \\ q_2 \\ q_3 \end{pmatrix} = \Lambda(u) \begin{pmatrix} u_1 \\ u_2 \\ u_3 \\ u_4 \end{pmatrix} + \begin{pmatrix} \mu_2 \\ 0 \\ 0 \end{pmatrix} \ . \tag{10}$$

Eq. (9) suggests to define

$$\begin{pmatrix} p_1 \\ p_2 \\ p_3 \end{pmatrix} = \frac{1}{(u_1^2 + u_2^2 + u_3^2 + u_4^2)} \Lambda(u) \begin{pmatrix} U_1 \\ U_2 \\ U_3 \\ U_4 \end{pmatrix} . \tag{11}$$

We shall use the notation $u_0 = t$, $U_0 = T$, where T is the variable conjugated to t. The aim of this section is to verify that (10) and (11) define a canonical transformation (see [7] for further details). Recall that

$$r_1 = \sqrt{(x - \mu_2)^2 + y^2 + z^2} = u_1^2 + u_2^2 + u_3^2 + u_4^2 = |u|^2 .$$

Defining the scalar product

$$l(U, u) \equiv (U, u^*)$$

with $u^* = (u_4, -u_3, u_2, -u_1)$ (last row of $A(u)$), one obtains

$$\begin{pmatrix} p_1 \\ p_2 \\ p_3 \\ \rho \end{pmatrix} = \frac{1}{2r_1} A(u) \begin{pmatrix} U_1 \\ U_2 \\ U_3 \\ U_4 \end{pmatrix} ,$$

where the auxiliary variable ρ is defined as

$$\rho = \frac{1}{2r_1}(U_1 u_4 - U_2 u_3 + U_3 u_2 - U_4 u_1) = \frac{1}{2r_1} \, l(U, u) .$$

Since

$$\begin{pmatrix} p_1 \\ p_2 \\ p_3 \\ \rho \end{pmatrix} (p_1, p_2, p_3, \rho) = \frac{1}{4r_1^2} r_1 |U|^2 = \frac{1}{4r_1} |U|^2$$

(recall that $A(u)A(u)^+ = r_1$), one obtains $(p_1^2 + p_2^2 + p_3^2 + \rho^2)r_1 = \frac{1}{4}|U|^2$, namely

$$(p_1^2 + p_2^2 + p_3^2)r_1 = \frac{1}{4}|U|^2 - \frac{1}{4r_1} l^2(U, u) . \tag{12}$$

We rewrite the Hamiltonian describing the spatial case in a fixed reference frame, evidentiating the term $\frac{1}{r_1}$ due to the interaction with $\overline{P}_1$; in the extended phase space, one has

$$H = \frac{1}{2}(p_1^2 + p_2^2 + p_3^2) + T - \frac{1}{r_1} - V(q_1, q_2, q_3, t) ;$$

introducing the fictitious time defined by $dt = r_1 ds$ one obtains

$$\tilde{H} = \frac{1}{2}(p_1^2 + p_2^2 + p_3^2)r_1 + Tr_1 - 1 - r_1 V(x, y, z, t) .$$

Using $r_1 = |u|^2$ and (12), one gets the Hamiltonian

$$\overline{H} = \frac{1}{8}|U|^2 + T|u|^2 - 1 + |u|^2 V(u, u_0) - \frac{1}{8|u|^2} l^2(U, u) . \qquad (13)$$

Lemma 1: The bilinear quantity

$$l(U, u) = U_1 u_4 - U_2 u_3 + U_3 u_2 - U_4 u_1$$

is a *first integral* of the new canonical equations

$$\frac{du_k}{ds} = \frac{\partial \overline{H}}{\partial U_k} , \qquad \frac{dU_k}{ds} = -\frac{\partial \overline{H}}{\partial u_k} \qquad (k = 0, 1, 2, 3, 4) ,$$

namely l is constant along the solutions of the equations of motion.
Proof: One has:

$$\frac{dl}{ds} = \sum_{j=1}^{4} \frac{\partial l}{\partial U_j} \frac{dU_j}{ds} + \frac{\partial l}{\partial u_j} \frac{du_j}{ds} = (u^*, \frac{dU}{ds}) - (U^*, \frac{du}{ds})$$

and denoting by $U_0 \equiv T$, the new canonical equations are

$$\frac{du}{ds} = \frac{\partial \overline{H}}{\partial U} = \frac{1}{4}U - \frac{1}{4|u|^2}u^* l(U, u)$$

$$\frac{dU}{ds} = -\frac{\partial \overline{H}}{\partial u} = -2U_0 u - \frac{\partial}{\partial u}(|u|^2 V) - \frac{1}{4|u|^4}u l^2(U, u) - \frac{1}{4|u|^2}U^* l(U, u) .$$

Observing that $(u^*, u) = (U^*, U) = 0$, one has

$$\frac{dl}{ds} = -2U_0(u^*, u) - (u^*, \frac{\partial}{\partial u}(|u|^2 V)) - \frac{1}{4|u|^4}(u^*, u) l^2(U, u)$$

$$- \frac{1}{4|u|^4}(u^*, U^*) l(U, u) - \frac{1}{4}(U^*, U) + \frac{1}{4|u|^2}(U^*, u^*) l(U, u)$$

$$= -(u^*, \frac{\partial}{\partial u}[(u, u)V]) = -2(u^*, u)V - |u|^2(u^*, \frac{\partial V}{\partial u}) = -|u|^2(u^*, \frac{\partial V}{\partial u}) .$$

From direct computations it follows that $\frac{\partial V}{\partial u} = 2A^+(u)\frac{\partial V}{\partial q_1}$; moreover, one has
that

$$(u^*, \frac{|u|^2}{2}A^+(u)(-\frac{\partial V}{\partial q_1})) = 0 .$$

Indeed, if $y \equiv -\frac{\partial V}{\partial x}$ and $v \equiv 2|u|^2 A^+(u)y$, then $(u^*, v) = (v, u^*) = 0$. In fact,
$(v, u^*) = l(v, u) = u_4 v_1 - u_3 v_2 + u_2 v_3 - u_1 v_4$ is equal to the fourth component of

$$A(u)v = 2A(u)A^+(u)y |u|^2 = 2(u, u)y |u|^2 ,$$

which is zero since it coincides with the fourth component of $y \equiv -\frac{\partial V}{\partial q_1}$.

Finally we have $(u^*, \frac{\partial V}{\partial u}) = 0$, which implies $\frac{dl}{ds} = 0$ along the equations of motions.

Lemma 2: Assume that the initial values $u(0)$, $U(0)$ of $u(s)$, $U(s)$ satisfy at $s = 0$ the bilinear relation

$$l(U(0), u(0)) = U_1(0)u_4(0) - U_2(0)u_3(0) + U_3(0)u_2(0) - U_4(0)u_1(0) = 0 .$$

Then one has:
a) the value of the first integral l is zero: $l(U, u) = U_1u_4 - U_2u_3 + U_3u_2 - U_4u_1 = 0$;
b) the Hamiltonian (13) is equivalent to the reduced Hamiltonian

$$\hat{H} = \frac{1}{8}|U|^2 + U_0|u|^2 - 1 + |u|^2 V(u, u_0) .$$

Proof:
a) It is a corollary of Lemma 1.
b) The canonical equations related to the Hamiltonian $\overline{H}$ and $\hat{H}$ possess the same solutions as a consequence of a), since the term $l^2(U, u)$ can be factored out in $\overline{H} - \hat{H}$ and differentiation leaves a factor l, which is zero along the given solutions.

At this stage we need the fact that to each set of initial conditions in the physical space $q_1(0)$, $q_2(0)$, $q_3(0)$, $p_1(0)$, $p_2(0)$, $p_3(0)$ we can associate a set of initial conditions in the parametric space $u(0)$, $u_0(0)$, $U(0)$, $U_0(0)$ (obtained through (10), (11)), such that

(i) $u_0(0) = 0$, $U_0(0) = -H(q_1(0), q_2(0), q_3(0), p_1(0), p_2(0), p_3(0), 0)$;

(ii) $l(U(0), u(0)) = U_1(0)u_4(0) - U_2(0)u_3(0) + U_3(0)u_2(0) - U_4(0)u_1(0) = 0$.

Notice that to obtain (ii) one can proceed as at the end of Lemma 1, setting $v = A^+(u)y$ with $y = q_1'(0)$ and defining $U(0) \equiv \frac{1}{2|u(0)|^2} A^+(u(0))q_1'(0)$.

Main Theorem: Suppose that $u(0)$, $u_0(0)$, $U(0)$, $U_0(0)$ satisfy (i) and (ii); then the solutions of

$$\frac{du_k}{ds} = \frac{\partial \hat{H}}{\partial U_k} , \qquad \frac{dU_k}{ds} = -\frac{\partial \hat{H}}{\partial u_k} \qquad (k = 0, 1, 2, 3, 4)$$

give by (10) and (11) the solution of Hamilton's equations relative to H with the corresponding initial data:

$$\frac{dQ}{dt} = \frac{\partial H}{\partial P} , \qquad \frac{dP}{dt} = -\frac{\partial H}{\partial Q} \qquad (Q = (q_1, q_2, q_3) , \quad P = (p_1, p_2, p_3)) .$$

Proof: To prove the canonicity, we use the criterion based on Poisson brackets, namely we need to check that

$$(a) \qquad \{q_k, q_l\} = \sum_{j=0}^{4} \left(\frac{\partial q_k}{\partial U_j} \frac{\partial q_l}{\partial u_j} - \frac{\partial q_k}{\partial u_j} \frac{\partial q_l}{\partial U_j} \right) = 0$$

$$(b) \qquad \{p_k, q_l\} = \sum_{j=0}^{4} \left(\frac{\partial p_k}{\partial U_j} \frac{\partial q_l}{\partial u_j} - \frac{\partial p_k}{\partial u_j} \frac{\partial q_l}{\partial U_j} \right) = \delta_{kl}$$

$$(c) \qquad \{p_k, p_l\} = \sum_{j=0}^{4} \left(\frac{\partial p_k}{\partial U_j} \frac{\partial p_l}{\partial u_j} - \frac{\partial p_k}{\partial u_j} \frac{\partial p_l}{\partial U_j} \right) = 0 \ ,$$

for $k = 0, 1, 2, 3$, $l = 0, 1, 2, 3$ and for every s. The case $k = 0$ or $l = 0$ is trivial; therefore the sum can be restricted to $j = 1, 2, 3, 4$. Since the variables U do not appear in the KS–transformation, one has $\frac{\partial q_k}{\partial U_j} = 0$, namely $\{q_k, q_l\} = 0$, which implies (a), while (b) becomes

$$\{p_k, q_l\} = \sum_{j=1}^{4} \frac{\partial p_k}{\partial U_j} \frac{\partial q_l}{\partial u_j} = \delta_{kl} \qquad (k, l = 1, 2, 3) \ . \tag{14}$$

Using matrix notation one has

$$\left(\frac{\partial q_l}{\partial u_j} \right) = 2\Lambda(u) \ , \qquad\qquad \left(\frac{\partial p_k}{\partial U_j} \right) = \frac{1}{2|u|^2} \Lambda(u) \ ,$$

$$\left(\frac{\partial p_k}{\partial u_j} \right) = -\frac{1}{|u|^4} \Lambda(u) \begin{pmatrix} U_1 \\ U_2 \\ U_3 \\ U_4 \end{pmatrix} (u_1, u_2, u_3, u_4) + \frac{1}{2|u|^2} \Lambda(U) \ ,$$

for $l, k = 1, 2, 3$, $j = 1, 2, 3, 4$. Therefore, (14) is equivalent to

$$\{p_k, q_l\} = \frac{1}{|u|^2} [\Lambda(u)\Lambda^+(u)]_{kl} = \frac{1}{|u|^2} (u, u) \ (Id.)_{kl} = \delta_{kl} \ ,$$

due to the orthogonality of Λ. It remains to prove (c) whose proof makes use of the bilinear relation, which is valid along the solution due to $a)$ of Lemma 2. Condition (c) can be checked by lenghty computations. Alternatively, due to the symmetry of the two terms in (c), it is convenient to prove (c) by showing that the matrix $A - B$ is symmetric, where

$$A - B = \left(\sum_{j=1}^{4} \frac{\partial p_k}{\partial u_j} \frac{\partial p_l}{\partial U_j} \right) 2|u|^6$$

$$A \quad \equiv \frac{1}{2} |u|^2 \Lambda(U)\Lambda^+(U)$$

$$B \quad \equiv \Lambda(U) \begin{pmatrix} U_1 \\ U_2 \\ U_3 \\ U_4 \end{pmatrix} (u_1, u_2, u_3, \dot{u}_4)\Lambda^+(U)$$

$$\quad = \Lambda(U) \begin{pmatrix} U_1 \\ U_2 \\ U_3 \\ U_4 \end{pmatrix} \left[\Lambda(U) \begin{pmatrix} u_1 \\ u_2 \\ u_3 \\ u_4 \end{pmatrix} \right]^+ \ .$$

We refer to [7] for the explicit proof of the symmetry of $A - B$.

5   The radial–inversion transformation

In this section we review a regularization technique which is alternative to the Levi–Civita and KS transformations. This method, to which we refer as the *radial–inversion transformation*, is developed in all details in [5].

Consider three bodies P_1, P_2, P_3 with masses m_1, m_2, m_3; let $(q_1, ..., q_9)$ be their spatial coordinates in a fixed reference frame and let $(p_1, ..., p_9)$ be the corresponding momenta. We denote by $(x_1, ..., x_6)$ the coordinates of P_1 and P_2 relative to P_3, i.e.

$$x_k = q_k - q_{k+6} , \qquad x_{k+3} = q_{k+3} - q_{k+6} ,$$
$$x_{k+6} = q_{k+6} , \qquad k = 1, 2, 3$$

(notice that x_7, x_8, x_9 still denote the coordinates of P_3). The corresponding momenta become

$$y_k = p_k , \qquad y_{k+3} = p_{k+3} , \qquad y_{k+6} = p_k + p_{k+3} + p_{k+6} , \qquad k = 1, 2, 3 .$$

Therefore, the equations of motion are given by

$$\dot{x}_k = H_{y_k} , \qquad \dot{y}_k = -H_{x_k} , \qquad k = 1, ..., 9 ,$$

where $H = T - U$ with

$$T = \frac{1}{2} \sum_{k=1}^{3} \left(\frac{y_k^2}{m_1} + \frac{y_{k+3}^2}{m_2} + \frac{(y_{k+6} - y_k - y_{k+3})^2}{m_3} \right)$$

and

$$U = \frac{m_1 m_3}{(x_1^2 + x_2^2 + x_3^2)^{1/2}} + \frac{m_2 m_3}{(x_4^2 + x_5^2 + x_6^2)^{1/2}} + \frac{m_1 m_2}{\left[(x_1 - x_4)^2 + (x_2 - x_5)^2 + (x_3 - x_6)^2 \right]^{1/2}} .$$

We remark that the Hamiltonian function does not depend on x_7, x_8, x_9, so that y_7, y_8, y_9 are constants of motion corresponding to the integrals of the center of mass.

Without loss of generality, we can assume that the origin of the reference frame coincides with the center of mass, so that

$$x_{k+6} = -\frac{m_1 x_k + m_2 x_{k+3}}{M} , \qquad k = 1, 2, 3 ,$$

where $M \equiv m_1 + m_2 + m_3$ is the total mass. Consequently, the kinetic energy becomes

$$T = \frac{1}{2} \left(\frac{1}{m_1} + \frac{1}{m_3} \right) \sum_{k=1}^{3} y_k^2 + \frac{1}{2} \left(\frac{1}{m_2} + \frac{1}{m_3} \right) \sum_{k=1}^{3} y_{k+3}^2 + \frac{1}{m_3} \sum_{k=1}^{3} y_k y_{k+3} .$$

Let us consider the case of a collision between P_1 and P_3, assuming that such event takes place at time $t = t_c$. To simplify the notation, we define

$$x^2 \equiv x_1^2 + x_2^2 + x_3^2 , \qquad y^2 \equiv y_1^2 + y_2^2 + y_3^2 ;$$

a collision occurs whenever $x \to 0$ as $t \to t_c$.

We introduce a fictitious time s defined through the relation

$$dt = x \, ds ;$$

the equations of motion with respect to the new time are

$$x_k' = x H_{y_k} , \qquad y_k' = -x H_{x_k} , \qquad k = 1, ..., 6 ,$$

where the prime denotes derivative with respect to s. Since the above equations are no longer in Hamiltonian form, we take advantage from the conservation of the energy, as follows. Assume that $H = h$ for a fixed value of the energy h and replace the Hamiltonian function by

$$\tilde{H} = x(H - h) = xT - xU - xh ; \tag{15}$$

therefore the dynamics of the solutions for which $H = h$ (i.e., $\tilde{H} = 0$) is governed by the equations

$$x_k' = \tilde{H}_{y_k} , \qquad y_k' = -\tilde{H}_{x_k} , \qquad k = 1, ..., 6 .$$

Let us start with the study of the two-body problem obtained ignoring the existence of P_2. Then, the kinetic energy and the force function reduce to

$$T = \frac{1}{2}\left(\frac{1}{m_1} + \frac{1}{m_3}\right) y^2$$
$$U = \frac{m_1 m_3}{x} ,$$

so that, neglecting constant terms, the Hamiltonian can be written as

$$\tilde{H} = \frac{1}{2}\left(\frac{1}{m_1} + \frac{1}{m_3}\right) xy^2 - hx .$$

Restricting to zero energy (i.e., taking $h = 0$) and normalizing the masses so that $\frac{1}{2}\left(\frac{1}{m_1} + \frac{1}{m_3}\right) = 1$, one obtains

$$\tilde{H} = xy^2 = (x_1^2 + x_2^2 + x_3^2)^{\frac{1}{2}} (y_1^2 + y_2^2 + y_3^2)$$

with associated Hamilton's equations

$$x_k' = \tilde{H}_{y_k} , \qquad y_k' = -\tilde{H}_{x_k} , \qquad k = 1, 2, 3 .$$

From Hamilton–Jacoby theory, since $\tilde{H}$ does not depend on s, the solution can be obtained looking for a function w of the form

$$w(x_k, \xi_k, s) = v(x_k, \xi_k) - \lambda(\xi_k)s ,$$

such that

$$\tilde{H}(x_k, v_{x_k}) = \lambda(\xi_k) . \tag{16}$$

The equations associated to the generating function are

$$y_k = v_{x_k}$$
$$\eta_k = \lambda_{\xi_k} s - v_{\xi_k} , \qquad k = 1, 2, 3 . \tag{17}$$

Instead of (17) we consider the following canonical transformation independent on s:

$$y_k = v_{x_k}$$
$$\eta_k = -v_{\xi_k} , \qquad k = 1, 2, 3 . \tag{18}$$

Let us now start from the discussion of the *planar case* for which (16) becomes

$$(x_1^2 + x_2^2)^{\frac{1}{2}} (v_{x_1}^2 + v_{x_2}^2) = \lambda(\xi_k) .$$

To this end, let $z = x_1 + ix_2$; we look for v as the imaginary part of an analytic function $f(z) = u + iv$. By Cauchy–Riemann equations, one finds

$$u_{x_1} = v_{x_2} , \qquad v_{x_1}^2 + v_{x_2}^2 = u_{x_1}^2 + v_{x_1}^2 = |f_z|^2 ,$$

from which it follows that $|zf_z^2| = \lambda(\xi_k)$ is constant in z. Let $\zeta = \xi_1 + i\xi_2$ be a complex constant and $zf_z^2 = \bar{\zeta}$. Then, one finds that $f_z(z) = 2(\bar{\zeta}/z)^{\frac{1}{2}}$, which gives by integration: $f(z) = 2\sqrt{\bar{\zeta}z}$. From this solution, it follows that

$$v^2 = 2|\zeta z| - \bar{\zeta}z - \zeta\bar{z} = 2\left(\xi x - \sum_{k=1}^{2} \xi_k x_k\right) .$$

The generalization to the *spatial case* is immediately given by the following formulae:

$$v^2 = 2\left(\xi x - \sum_{k=1}^{3} \xi_k x_k\right)$$
$$\xi = \sqrt{\xi_1^2 + \xi_2^2 + \xi_3^2}$$
$$x = \sqrt{x_1^2 + x_2^2 + x_3^2} ,$$

with associated Hamiltonian function

$$\tilde{H} = xy^2 = (x_1^2 + x_2^2 + x_3^2)^{\frac{1}{2}} (y_1^2 + y_2^2 + y_3^2) .$$

From the above relations one easily finds that

$$xvv_{x_k} = x_k\xi - \xi_k x$$
$$\xi vv_{\xi_k} = \xi_k x - x_k\xi ,$$

from which it follows that $xv_{x_k} = -\xi v_{\xi_k}$. Therefore, by (18) one obtains

$$xy_k = \xi\eta_k , \qquad k = 1,2,3 . \tag{19}$$

Moreover, from (18) and

$$x\sum_{k=1}^{3} v_{x_k}^2 = \xi , \qquad \xi\sum_{k=1}^{3} v_{\xi_k}^2 = x ,$$

one obtains

$$xy^2 = \xi , \qquad \xi\eta^2 = x .$$

Finally, the *radial–inversion transformation* is obtained through (19) as

$$\eta_k = \frac{y_k}{y^2}$$
$$y_k = \frac{\eta_k}{\eta^2} , \qquad k = 1,2,3 .$$

Moreover, one finds that (see [5])

$$x_k = \xi_k\eta^2 - 2\eta_k\sum_{l=1}^{3} \xi_l\eta_l$$
$$\xi_k = x_ky^2 - 2y_k\sum_{l=1}^{3} x_ly_l , \qquad k = 1,2,3 .$$

Let us now come back to the 3–body problem for which we rewrite the radial–inversion transformation as

$$\eta_k = \frac{y_k}{y^2} , \qquad \xi_k = x_ky^2 - 2y_k\sum_{l=1}^{3} x_ly_l , \qquad k = 1,2,3$$
$$\eta_k = y_k , \qquad \xi_k = x_k , \qquad\qquad\qquad k = 4,5,6 .$$

Hamilton's equations in the new variables are

$$\xi_k' = \hat{H}_{\eta_k}$$
$$\eta_k' = -\hat{H}_{\xi_k} , \qquad k = 1,...,6 ,$$

where $\hat{H}$ denotes the Hamiltonian $\tilde{H}$ expressed in the new set of variables (ξ_k, η_k). Recalling (15) (for a nonzero value of h) and using the following relations

$$xy^2 = \xi$$
$$x = \xi\eta^2$$

$$x_k = \xi_k\eta^2 - 2\eta_k \sum_{l=1}^{3} \xi_l\eta_l$$

$$y_k = \frac{\eta_k}{\eta^2} \qquad\qquad k = 1, 2, 3 \,,$$

one easily finds that

$$xT = \frac{1}{2}\left(\frac{1}{m_1} + \frac{1}{m_3}\right)\xi + \frac{1}{2}\left(\frac{1}{m_2} + \frac{1}{m_3}\right)\xi\eta^2 \sum_{k=1}^{3} \eta_{k+3}^2 + \frac{1}{m_3}\xi \sum_{k=1}^{3} \eta_k\eta_{k+3}$$

$$xU = m_1m_3 + m_2\xi\eta^2 \left(\frac{m_3}{r_{23}} + \frac{m_2}{r_{12}}\right) \tag{20}$$

$$xh = \xi\eta^2 h \,,$$

where

$$r_{23}^2 = \sum_{k=1}^{3} \xi_{k+3}^2 \,, \qquad r_{12}^2 = \sum_{k=1}^{3} (x_k - \xi_{k+3})^2 \,.$$

At collision, i.e. when P_1 tends to P_3, the distances r_{12} and r_{23} have positive limits and therefore the equations of motion in the (ξ, η) variables are regular. This concludes the regularizing transformation by radial–inversion.

References

1. G.D. Birkhoff: Trans. Am. Math. Soc. **18**, 199 (1917)
2. H. Goldstein: *Classical Mechanics.* (Addison–Wesley 1980)
3. A. Hurwitz: Math. Werke **II**, 565 (1933)
4. P. Kustaanheimo, E. Stiefel: J. Reine Angew. Math **218**, 204 (1965)
5. C.L. Siegel, J.K. Moser: *Lectures on Celestial Mechanics.* (Springer–Verlag, Berlin, Heidelberg, New York 1971)
6. E.L. Stiefel, M. Rössler, J. Waldvogel, C.A. Burdet: NASA Contractor Report **CR-769**, Washington, 1967
7. E.L. Stiefel, G. Scheifele: *Linear and Regular Celestial Mechanics.* (Springer–Verlag, Berlin, Heidelberg, New York 1971)
8. E.L. Stiefel, J. Waldvogel: C. R. Acad. Sc. Paris **260**, 805 (1965)
9. V. Szebehely: *Theory of Orbits.* (Academic Press, New York and London 1967)
10. J. Waldvogel: Bull. Astron. **3**, II, 2, 295 (1967)

The Birkhoff and B_3 Regularizing Transformations

Maria Gabriella Della Penna

Observatoire de Nice, B.P.4229, F-06304 Nice Cedex 4 (France)

Abstract. The Birkhoff transformation for the restricted circular planar three body problem is introduced using a Hamiltonian formalism. A generalization of the Birkhoff transformation (hereafter the B_3 transformation) is derived in order to regularize the circular spatial three body problem by means of a geometrical approach. These transformations have the important property to provide a simoultaneous regularization of two singularities.
Some results about numerical integrations before and after the regularization are presented.

1 Introduction

In this section we focus our attention on the Birkhoff regularizing method applied to the restricted circular planar three body problem. We next introduce the B_3 regularization, namely a generalization of the Birkhoff regularization for 3–dimensional motions.

The Levi–Civita and KS transformations were introduced in a previous chapter in order to regularize a single collision in the two body problem, or in the restricted problem. However when dealing with the restricted circular problem (i.e. the motion of a satellite with negligible mass in the gravitational field generated by the primaries), Newton's equations are characterized by two singularities at the attracting centers and sometimes a simultaneous regularization of both singularities is needed. This can be performed by using the Birkhoff and B_3 transformations.

The first section of this paper is devoted to the Birkhoff transformation for the restricted circular planar three body problem; in the second part, we introduce the B_3 regularization for the spatial problem ([4], [7], [6], [8]). Finally we show how the Birkhoff regularizing transformation works, performing numerical integrations for some initial conditions of the pre-regularized and post-regularized equations of motion of the restricted circular planar three body problem ([2]).

2 The Birkhoff regularization

Let $\bar{P}_1$, $\bar{P}_2$ be two massive bodies which interact each other in a Newtonian gravity field, and $\bar{P}_3$ a third body of negligible mass in the force field of the primaries. We suppose that $\bar{P}_1$ and $\bar{P}_2$ evolve in circular orbits around their center of mass. A synodic coordinate system (y_1, y_2) rotating with unit angular velocity is introduced, so that $\bar{P}_1$ and $\bar{P}_2$ are fixed on the y_1–axis, the origin being their center of gravity. We restrict the motion to planar orbits on the y–plane, so that we are dealing with a restricted–circular–planar three body problem. By a convenient choice of the units of measure, we normalize to one: the total mass of $\bar{P}_1$ and $\bar{P}_2$, their distance, and the gravitational costant. Denoting the mass of $\bar{P}_2$ by μ, its position is at $(1 - \mu, 0)$ in the y–plane, while $\bar{P}_1$ is located at $(-\mu, 0)$. Finally we denote by r_1, r_2 the distances of the moving particle $\bar{P}_3$ from $\bar{P}_1$ and $\bar{P}_2$ respectively.

Under these hypotheses, the Hamiltonian function which describes the motion of $\bar{P}_3$ is given by:

$$H(y_1, y_2, \tilde{p}_1, \tilde{p}_2) = \frac{1}{2}(\tilde{p}_1{}^2 + \tilde{p}_2{}^2) + y_2 \tilde{p}_1 - y_1 \tilde{p}_2 - \frac{1 - \mu}{r_1} - \frac{\mu}{r_2}, \qquad (1)$$

where y_1, y_2 are the coordinates of $\bar{P}_3$, $\tilde{p}_1, \tilde{p}_2$ their impulses, and:

$$r_1 = \sqrt{(y_1 + \mu)^2 + y_2{}^2}, \quad r_2 = \sqrt{(y_1 - 1 + \mu)^2 + y_2{}^2}.$$

The Hamiltonian (1) has two singularities with respect to $\bar{P}_1$ and $\bar{P}_2$, and a simultaneous regularization at both attracting centers is needed. This is performed by the so–called Birkhoff's transformation, which maps the (y_1, y_2) physical plane into the parametric (Q_1, Q_2)-plane so that the Hamiltonian (1) has no longer singularities at the attracting centers.

Instead of using the mass center of the primaries as the origin of the coordinate system, we select the midpoint between the primaries as the origin. Let $q_1 + iq_2 = y_1 + iy_2 - \frac{1}{2} + \mu$ be the transformation which locates the primaries at $(q_1, q_2) = (\pm \frac{1}{2}, 0)$. We introduce a generating function:

$$W(q_1, q_2, \tilde{p}_1, \tilde{p}_2) = (q_1 + \frac{1}{2} - \mu)\tilde{p}_1 + q_2 \tilde{p}_2,$$

which depends from the old impulses and the new coordinates. Therefore we have:

$$y_1 = \frac{\partial W}{\partial \tilde{p}_1} = q_1 + \frac{1}{2} - \mu, \quad y_2 = \frac{\partial W}{\partial \tilde{p}_2} = q_2$$

$$p_1 = \frac{\partial W}{\partial q_1} = \tilde{p}_1, \qquad p_2 = \frac{\partial W}{\partial q_2} = \tilde{p}_2.$$

The new Hamiltonian becomes:

$$H_1(q_1, q_2, p_1, p_2) = \frac{1}{2}(p_1{}^2 + p_2{}^2) + q_2 p_1 - (q_1 + \frac{1}{2} - \mu)p_2 - \frac{1 - \mu}{r_1} - \frac{\mu}{r_2}, \qquad (2)$$

where:

$$r_1 = \sqrt{(q_1 + \frac{1}{2})^2 + q_2{}^2}, \quad r_2 = \sqrt{(q_1 - \frac{1}{2})^2 + q_2{}^2}.$$

The singularities are now located at $\bar{P}_1 = (-\frac{1}{2}, 0)$, $\bar{P}_2 = (\frac{1}{2}, 0)$.

The equations of motion in complex form become:

$$\ddot{q} + 2i\dot{q} = \nabla_q U(q),\qquad\qquad(3)$$

where $\nabla_q U(q)$ denotes the gradient of $U(q)$ with respect to q, $U(q) = \Omega(q) - \dfrac{C}{2}$ for a suitable constant C, and:

$$\Omega(q) = \frac{1}{2}[(1-\mu)r_1{}^2 + \mu r_2{}^2] + \frac{1-\mu}{r_1} + \frac{\mu}{r_2} = \frac{1}{2}[(1-\mu)r_1{}^2 + \mu r_2{}^2] + \Omega_c(q),$$

where $\Omega_c(q)$ is caleld the critical part of $\Omega(q)$,

$$\Omega_c(q) = \frac{1-\mu}{r_1} + \frac{\mu}{r_2}.$$

The Jacobi integral is given by the expression:

$$|\dot{q}|^2 = 2\Omega(q) - C = 2U(q).$$

Using complex notation, let $w = Q_1 + iQ_1$ (i is the imaginary part); the regularizing function is defined by:

$$q = h(w) = \alpha w + \frac{\beta}{w},$$

α, β are constants to be determined.

Beside the above transformation of coordinates, we define a time transformation

$$\frac{dt}{d\tau} = g(w) \equiv |k(w)|^2 = k(w)\overline{k}(w),$$

were $k(w)$ is a suitable complex function that will be defined later on, the bar denotes the conjugate, and t is the old time, wheareas τ is the new time.

We derive the equation of motion (3) in terms of the new variables, in order to point out more easily what is the singular part of the equations, and consequently we determine the constants α, β. One finds:

$$\dot{q} = \frac{dq}{dt} = \frac{dh}{dw}\frac{dw}{d\tau}\frac{d\tau}{dt} = h'w'\dot{\tau},$$

while:

$$\ddot{q} = h'w'\ddot{\tau} + (h''w'^2 + h'w'')\dot{\tau}^2.$$

The gradient operator becomes:

$$\overline{h'}\nabla_q U = \nabla_w U.$$

Therefore, the equations of motion (3) take the form:

$$w'' + 2i\frac{w'}{\dot\tau} + w'\frac{\ddot\tau}{\dot\tau^2} + w'^2\frac{h''}{h'} = \nabla_w U \frac{1}{|h'|^2\dot\tau^2}.$$
(4)

We know also that: $\dot\tau = \dfrac{1}{g} = \dfrac{1}{k\overline{k}}$, so that $\ddot\tau = -\dfrac{\dot g}{g^2}$, namely: $\dfrac{\ddot\tau}{\dot\tau^2} = -\dot g$. Since:

$$-\dot g = -(k\frac{\overline{dk}}{dw}\frac{dw}{d\tau} + \overline{k}\frac{dk}{dw}\frac{dw}{d\tau})\dot\tau = -(\frac{\overline{k'}w'}{\overline{k}} + \frac{k'w'}{k}),$$

equation (4) becomes:

$$w'' + 2ik\overline{k}w' - \frac{|w'|^2}{\overline{k}}\frac{d\overline{k}}{dw} + w'^2(\frac{h''}{h'} - \frac{k'}{k}) = \frac{|k|^4}{|h'|^2}\nabla_w U.$$
(5)

By the relation of the energy integral it follows that $|w'|^2 = \dfrac{|k|^4}{|h'|^2}2U$.
In fact, the Jacobi integral in terms of the old variables is:

$$|\dot q|^2 = 2\Omega(q) - C = 2U(q);$$
(6)

from (6) and

$$|\dot q| = |h'||w'|\frac{1}{|k|^2},$$

we get the relation of the energy integral in terms of the w–variables. Then, introducing

$$\frac{h''}{h'} - \frac{k'}{k} = \frac{d}{dw}(ln\frac{h'}{k}),$$

in (5) we get the equation:

$$w'' + 2ik\overline{k}w' + (w')^2\frac{d}{dw}(ln\frac{h'}{k}) = \frac{|k|^4}{|h'|^2}(2U\frac{dln\overline{k}}{dw} + \nabla_w U),$$

which may be written as:

$$w'' + 2ik\overline{k}w' = \nabla_w\frac{|k|^4}{|h'|^2}U - 2iw'Im(w'\frac{d}{dw}ln\frac{h'}{k}),$$

where the symbol Im stands for imaginary part. Finally, with the choice $k = h'$, the equation of motion assumes the form:

$$w'' + 2i|h'|^2w' = \nabla_w|h'|^2U.$$
(7)

At this point, we can determine the constants α and β by means of the following two conditions: it is required that the function h should eliminate both singularities, and that the points $\bar P_1$, $\bar P_2$ are fixed points of the transformation.

Concerning the first requirement we must consider the product of the critical part of $\Omega(w)$, say $\Omega_c(w)$, namely the part responsible for the singularities, with $|h'(w)|^2$, where:

$$\Omega_c(w) = \frac{1-\mu}{r_1} + \frac{\mu}{r_2} = \frac{1-\mu}{|\alpha w + \dfrac{\beta}{w} - \dfrac{1}{2}|} + \frac{\mu}{|\alpha w + \dfrac{\beta}{w} + \dfrac{1}{2}|},$$

$$|h'(w)|^2 = \frac{|\alpha w^2 - \beta|^2}{|w|^4};$$

this product can be written as:

$$\Omega_c(w) \cdot |h'(w)|^2 = \frac{1}{|w|^3}\left(\frac{(1-\mu)|\alpha w^2 - \beta|^2}{|\alpha w^2 + \beta + \dfrac{w}{2}|} + \frac{\mu|\alpha w^2 - \beta|^2}{|\alpha w^2 + \beta - \dfrac{w}{2}|}\right)$$

The singularity located at $q = \dfrac{1}{2}$ corresponds to the values for w:

$$w_{1,2} = \frac{1}{4\alpha}(1 \pm (1 - 16\alpha\beta)^{\frac{1}{2}})$$

which are the solutions of the equation:

$$\frac{1}{2} = \alpha w + \frac{\beta}{w}$$

The roots $w_{1,2}$ are the same roots of the denominator $|\alpha w^2 + \beta - \dfrac{w}{2}|$, and in order to eliminate the singularity, the roots of the corresponding numerator $|\alpha w^2 - \beta|$ must coincide with $w_{1,2}$, namely:

$$\frac{1}{4\alpha}(1 \pm (1 - 16\alpha\beta)^{\frac{1}{2}}) = \pm(\frac{\beta}{\alpha})^{\frac{1}{2}}. \tag{8}$$

Solving the equation (8) one gets:

$$\alpha\beta(16\alpha\beta - 1) = 0.$$

Since both α and β must be different form zero, one obtains $16\alpha\beta = 1$. This result shows that $w_1 = w_2 = \dfrac{1}{4\alpha}$, i.e. the image of $\bar{P}_2$ is uniquely determined by the transformation. If we look for a transformation which leaves $\bar{P}_2$ at $(\dfrac{1}{2}, 0)$, we must impose that $\dfrac{1}{4\alpha} = \dfrac{1}{2}$, namely $\alpha = \dfrac{1}{2}$ and $\beta = \dfrac{1}{8}$. The first term in the critical part $\Omega_c(w)$ has a singularity at $\bar{P}_1$ and its elimination is identical to the previous procedure; one finds the same values for α, β as in the regularization of $\bar{P}_2$.

The other singularity is due to the term $\dfrac{1}{|w|^3}$, but the singularity at $w = 0$ corresponds to $q \to \infty$, which does not have physical meaning in our problem. Finally, the Birkhoff transformation is given by

$$q = \frac{1}{2}(w + \frac{1}{4w}), \quad w = Q_1 + iQ_2,$$

namely:

$$q_1 = \frac{1}{2}(Q_1 + \frac{Q_1}{4(Q_1^2 + Q_2^2)})$$
$$q_2 = \frac{1}{2}(Q_2 - \frac{Q_2}{4(Q_1^2 + Q_2^2)}).$$

We next apply the Birkhoff transformation to the Hamiltonian (2) introducing the generating function

$$W(p_1, p_2, Q_1, Q_2) = p_1 f(Q_1, Q_2) + p_2 g(Q_1, Q_2),$$

where:

$$f(Q_1, Q_2) = \frac{1}{2}(Q_1 + \frac{Q_1}{4(Q_1^2 + Q_2^2)})$$
$$g(Q_1, Q_2) = \frac{1}{2}(Q_2 - \frac{Q_2}{4(Q_1^2 + Q_2^2)}).$$

The equations which relate the new and old variables are

$$q_1 = \frac{\partial W}{\partial p_1} = f(Q_1, Q_2), \qquad q_2 = \frac{\partial W}{\partial p_2} = g(Q_1, Q_2)$$
$$P_1 = \frac{\partial W}{\partial Q_1} = p_1 f_{Q_1} + p_2 g_{Q_1}, \quad P_2 = \frac{\partial W}{\partial Q_1} = p_1 f_{Q_2} + p_2 g_{Q_2},$$

with:

$$f_{Q_i} = \frac{\partial f}{\partial Q_i}, \quad g_{Q_i} = \frac{\partial g}{\partial Q_i}, \, i = 1, 2.$$

Moreover, the functions **f** and **g** are conjugate harmonic functions and therefore satisfy the Cauchy–Riemann relations:

$$f_{Q_1} = g_{Q_2}, \quad f_{Q_2} = -g_{Q_1}.$$

Let $\underline{P} = (P_1, P_2)$ and $\underline{p} = (p_1, p_2)$ be the vectors relative to the new and old impulses respectively. The change of the coordinates is given by: $\underline{P} = A\underline{p}$, where

$$A = \begin{pmatrix} f_{Q_1} & g_{Q_1} \\ f_{Q_2} & g_{Q_2} \end{pmatrix}.$$

Let:
$$D \equiv det(A) = (f_{Q_1})^2 + (g_{Q_1})^2 = (\frac{1}{2} + \frac{-Q_1^2 + Q_2^2}{8(Q_1^2 + Q_2^2)^2})^2 + (\frac{Q_1 Q_2}{4(Q_1^2 + Q_2^2)^2})^2.$$

Since $A^{-1} = \frac{A^+}{D}$ (A^+ is the transposed of the matrix A), we get the relation:

$p = A^+ \underline{P}/D$, and therefore: $\underline{p}^2 = \frac{P^2}{D}$, since A is an orthogonal matrix.
The quantities r_1, r_2 become

$$r_1 = \frac{\rho_1^2}{2\rho}, \quad r_2 = \frac{\rho_2^2}{2\rho}, \tag{9}$$

where

$$\rho_1 = ((Q_1 + \frac{1}{2})^2 + Q_2^2)^{\frac{1}{2}}, \quad \rho_2 = ((Q_1 - \frac{1}{2})^2 + Q_2^2)^{\frac{1}{2}},$$
$$\rho = (Q_1^2 + Q_2^2)^{\frac{1}{2}},$$

while $D = \frac{\rho_1^2 \rho_2^2}{4\rho^4}$.

In fact, $r_1 = |q + \frac{1}{2}| = |\frac{w}{2} + \frac{1}{8w} + \frac{1}{2}| = \frac{|w^2 + w + \frac{1}{4}|}{2|w|} = \frac{|w + \frac{1}{2}|^2}{2|w|} = \frac{\rho_1^2}{2\rho}$;

$r_2 = |q - \frac{1}{2}| = |\frac{w}{2} + \frac{1}{8w} - \frac{1}{2}| = \frac{|w^2 - w + \frac{1}{4}|}{2|w|} = \frac{|w - \frac{1}{2}|^2}{2|w|} = \frac{\rho_2^2}{2\rho}$;

The term: $q_2 p_1 - q_1 p_2 - (\frac{1}{2} - \mu)p_2$ is transformed into:

$$\frac{1}{2D}[P_1 \frac{\partial}{\partial Q_2}(f^2 + g^2) - P_2 \frac{\partial}{\partial Q_1}(f^2 + g^2) - (1 - 2\mu)(P_1 g_{Q_1} + P_2 f_{Q_1})]. \tag{10}$$

To prove (10), we first show that:

$$\frac{1}{2D}[P_1 \frac{\partial}{\partial Q_2}(f^2 + g^2) - P_2 \frac{\partial}{\partial Q_1}(f^2 + g^2)] = q_2 p_1 - p_2 q_1.$$

In fact, the l.h.s. is equal to:

$$\frac{1}{D}[P_1(f f_{Q_2} + g g_{Q_2}) - P_2(f f_{Q_1} + g g_{Q_1})]$$
$$= \frac{1}{D}[(p_1 f_{Q_1} + p_2 g_{Q_1})(f f_{Q_2} + g g_{Q_2}) - (p_1 f_{Q_2} + p_2 g_{Q_2})(f f_{Q_1} + g g_{Q_1})]$$
$$= \frac{1}{f_{Q_1}^2 + g_{Q_1}^2}[p_1 g(f_{Q_1}^2 + g_{Q_1}^2) - p_2 f(f_{Q_1}^2 + g_{Q_1}^2)]$$
$$= q_2 p_1 - p_2 q_1.$$

By the relation: $\underline{p} = A^+ \underline{P}/D$, we get:

$$p_1 = \frac{1}{D}[f_{Q_1} P_1 - g_{Q_1} P_2], \quad p_2 = \frac{1}{D}[g_{Q_1} P_1 + f_{Q_1} P_2]$$

Finally,

$$-(\frac{1}{2} - \mu)p_2 = \frac{1}{2D}[-(1 - 2\mu)(g_{Q_1}P_1 + f_{Q_1}P_2)],$$

and the relation (10) is proved.
Finally, the Hamiltonian in terms of the regularized variables is:

$$H_2(Q_1, Q_2, P_1, P_2) = \frac{1}{2D}[P_1{}^2 + P_2{}^2 + P_1\frac{\partial}{\partial Q_2}(f^2 + g^2)$$

$$-P_2\frac{\partial}{\partial Q_1}(f^2 + g^2) - (1 - 2\mu)(P_1g_{Q_1} + P_2f_{Q_1})] - (1 - \mu)\frac{2\rho}{\rho_1{}^2} - \mu\frac{2\rho}{\rho_2{}^2}.$$

Let $\Gamma = T + H_2(Q_1, Q_2, P_1, P_2)$ be the Hamiltonian in the extended phase space, where T is the variable conjugated to the time t. Now the time is a generalized coordinate, and we can perform also time transformations to the Hamiltonian Γ.

Introducing the fictitious time s by the relation $dt = Dds$, the new Hamiltonian becomes:

$$\Gamma^* = D\Gamma = DT + \frac{1}{2}[P_1{}^2 + P_2{}^2 + P_1\frac{\partial}{\partial Q_2}(f^2 + g^2)$$

$$-P_2\frac{\partial}{\partial Q_1}(f^2 + g^2) - (1 - 2\mu)(P_1g_{Q_1} + P_2f_{Q_1})] - D[(1 - \mu)\frac{2\rho}{\rho_1{}^2} + \mu\frac{2\rho}{\rho_2{}^2}].$$

The associated Hamilton's equations with respect to the fictitious time are:

$$Q_1' = P_1 + \frac{1}{2}[\frac{\partial}{\partial Q_2}(f^2 + g^2) - (1 - 2\mu)g_{Q_1}]$$

$$Q_2' = P_2 - \frac{1}{2}[\frac{\partial}{\partial Q_1}(f^2 + g^2) - (1 - 2\mu)f_{Q_1}]$$

$$t' = D$$

$$P_1' = -\frac{\partial\Gamma^*}{\partial Q_1} \tag{11}$$

$$P_2' = -\frac{\partial\Gamma^*}{\partial Q_2}$$

$$T' = 0.$$

We remark that the Hamiltonian Γ^* has no longer singularities. In fact, the singular term is:

$$D[(1 - \mu)\frac{2\rho}{\rho_1{}^2} + \mu\frac{2\rho}{\rho_2{}^2}] = \frac{1}{2\rho^3}[(1 - \mu)\rho_2{}^2 + \mu\rho_1{}^2]$$

and the only singularity of the whole Hamiltonian is at $\rho = 0$, corresponding to $r_1 \to \infty$, $r_2 \to \infty$, as we can check replacing $Q_1 = Q_2 = 0$ in (9). In the transformed Q–plane the origin becomes a new singular point. This, however, corresponds to infinity in the physical plane and we can conclude that all points of the finite physical plane have been regularized by the transformation.

In order to obtain the equations of motion (see (7)) one must evaluate the function $U|h'(w)|^2 = (\Omega - C/2)|h'(w)|^2$, which in our case becomes:

$$U|h'(w)|^2 = \frac{\rho_1^2 \rho_2^2}{32\rho^6}[(1-\mu)\rho_1^4 + \mu\rho_2^4] + \frac{1}{2\rho^3}[(1-\mu)\rho_2^2 + \mu\rho_1^2] - \frac{C\rho_1^2\rho_2^2}{8\rho^4}. \tag{12}$$

We observe that in (12) all points of the finite physical plane have been regularized by the transformation. Moreover relation (12) allows to express the velocity of the particle in the regularized coordinates, namely:

$$|\dot{Q}|^2 = 2U|h'(w)|^2. \tag{13}$$

At $\bar{P}_1$, $\rho_1 = 0$, $\rho_2 = 1$, $\rho = \dfrac{1}{2}$, and equation (13) gives $|\dot{Q}|^2 = 8(1 - \mu)$, so that $|\dot{Q}| = \sqrt{8(1-\mu)}$. Analogously, at $\bar{P}_2$ the velocity is $\sqrt{8\mu}$. We could expect such finite values for the velocities, since P_1 and P_2 have been regularized. By the relations (12), (13) it follows that the absolute value of the velocity in the regularized plane is determinated once the value of μ is fixed and once the initial conditions on the position, velocity vector and Jacobi integral is done. In particular, this property allows us to determine the value of the velocity at the primaries in the parametric plane. We remark that in the physical plane it is not possible to determine the value of the velocity at the primaries, since the equations of motion are singular at these points.

3   The B_3 regularization

The problem treated till now can be summarized as follows: two bodies having masses μ and $1 - \mu$ evolve around their center of mass in circular orbits. A third body of negligible mass moves on the orbital plane of the primaries and is subject to their gravitational field. A generalization of this problem is the so–called restricted–circular–spatial three body problem which allows the third particle to move in the space. The primaries $\bar{P}_1$, $\bar{P}_2$ still revolve in the same plane around their center of mass describing circular orbits. A synodic reference frame is introduced, so that the orbital plane of the primaries x, y rotates with unit mean motion; the angular velocity vector is normal to the plane along the z–axis, which forms a right–handed system with the coordinates x, y. Then, the primaries are fixed on the x–axis, with coordinates $\bar{P}_1 \equiv (-\mu, 0, 0)$, $\bar{P}_2 \equiv (1 - \mu, 0, 0)$. A third particle $\bar{P}_3$ of infinitesimal mass evolves in the space (x, y, z), not influencing the motion of the primaries, but subject to the Newtonian gravitational forces of $\bar{P}_1$, $\bar{P}_2$.

Denoting by (x, y, z) the coordinates of $\bar{P}_3$, the equations of motion describing the dynamics of $\bar{P}_3$ are:

$$\ddot{x} - 2\dot{y} = \Omega_x$$
$$\ddot{y} + 2\dot{x} = \Omega_y$$
$$\ddot{z} + z = \Omega_z,$$

where

$$\Omega(x,y,z) = \frac{1}{2}(x^2 + y^2 + z^2) + \frac{1-\mu}{r_1} + \frac{\mu}{r_2},$$

and

$$r_1 = \sqrt{(x+\mu)^2 + y^2 + z^2}, \quad r_2 = \sqrt{(x-1+\mu)^2 + y^2 + z^2}.$$

The Hamiltonian function describing the spatial problem in the synodic reference frame is given by:

$$H(P_1, P_2, P_3, Q_1, Q_2, Q_3) = \frac{1}{2}(P_1^2 + P_2^2 + P_3^2) + Q_2 P_1 - Q_1 P_2 - \frac{1-\mu}{r_1} - \frac{\mu}{r_2}, \quad (14)$$

where: $r_1 = \sqrt{(Q_1 + \mu)^2 + Q_2^2 + Q_3^2}$, $r_2 = \sqrt{(Q_1 - 1 + \mu)^2 + Q_2^2 + Q_3^2}$.

Also in the spatial case we are dealing with a Hamiltonian which presents two singularities at both attracting centers, corresponding to $r_1 = 0$, and $r_2 = 0$. In order to regularize such singularities we can perform a generalization of the Birkhoff transformation. In the spatial case, four generalized coordinates v_1, v_2, v_3, v_4 are introduced, the derivation of the Birkhoff transformation is more complicated.

In this section, we derive the Birkhoff transformation in a geometrical way (see [6], [8], [4]), which is complementary to the presentation done for the planar case.

We recall that $\bar{P}_1 \equiv (-\mu, 0, 0)$, $\bar{P}_2 \equiv (1 - \mu, 0, 0)$ have fixed locations on the x-axis. By means of a translation, we place $\bar{P}_1$ and $\bar{P}_2$ at $(-1, 0, 0)$ and $(1, 0, 0)$, respectively.

The transformation applied to the x–coordinate is:

$$x = \frac{1}{2}x_1 + \left(\frac{1}{2} - \mu\right),$$

where x_1 is the new abscissa, while the other coordinates remain unchanged.

Now, we perform an inversion, that is a transformation by reciprocal radii, having center at $(1, 0, 0)$ and radius equal to $\sqrt{2}$. The coordinates of $\bar{P}_1$, $\bar{P}_2$ are transformed following the rules:

$$x_2 - 1 = \frac{2(x_1 - 1)}{(x_1 - 1)^2 + y_1^2 + z_1^2}$$

$$y_2 = \frac{2y_1}{(x_1 - 1)^2 + y_1^2 + z_1^2}, \quad z_2 = \frac{2z_1}{(x_1 - 1)^2 + y_1^2 + z_1^2}.$$

Therefore, $\bar{P}_1$ is put at the origin of the (x_2, y_2, z_2) space, while $\bar{P}_2$ is ejected at infinity.

We next perform a KS–transformation, which has not only regularizing properties at the points of the finite physical plane (in this case the origin), but also at infinity, and allows us the regularization of both singularities. Instead of the co-ordinates (x_2, y_2, z_2) by means of the KS–transformation a 4–dimensional space u_1, u_2, u_3, u_4 is introduced as

$$x_2 = u_1^2 - u_2^2 - u_3^2 + u_4^2$$
$$y_2 = 2(u_1 u_2 - u_3 u_4)$$
$$z_2 = 2(u_1 u_3 + u_2 u_4).$$

Then we perform another inversion so that $\bar{P}_1$ is placed at $(-1, 0, 0)$ and $\bar{P}_2$ at $(1, 0, 0)$; let

$$u_1 - 1 = \frac{2(V_1 - 1)}{(V_1 - 1)^2 + V_2^2 + V_3^2 + V_4^2}$$
$$u_k = \frac{2V_k}{(V_1 - 1)^2 + V_2^2 + V_3^2 + V_4^2}, \qquad k = 2, 3, 4.$$

Finally we place $\bar{P}_1$ and $\bar{P}_2$ at $(-\frac{1}{2}, 0, 0)$ and $(\frac{1}{2}, 0, 0)$, respectively. The formulae which link the coordinates of the physical plane (x, y, z) to those of the parametric plane (v_1, v_2, v_3, v_4) are given by:

$$x = \frac{1}{2} - \mu + \frac{1}{2}[v_1 + \frac{v_1(v_4^2 + \frac{1}{4})}{v_1^2 + v_2^2 + v_3^2}]$$
$$y = \frac{1}{2}[v_2 + \frac{v_2(v_4^2 - \frac{1}{4}) - v_3 v_4}{v_1^2 + v_2^2 + v_3^2}]$$
$$z = \frac{1}{2}[v_3 + \frac{v_3(v_4^2 - \frac{1}{4}) + v_2 v_4}{v_1^2 + v_2^2 + v_3^2}].$$

Notice that if $v_3 = v_4 = 0$, then we get the planar Birkhoff transformation. Let s be the fictitious time introduced by $dt = D ds$, where:

$$D = \frac{r_1 r_2}{v_1^2 + v_2^2 + v_3^2},$$
$$r_1 = \frac{1}{2} \frac{(v_1 + \frac{1}{2})^2 + v_2^2 + v_3^2 + v_4^2}{\sqrt{v_1^2 + v_2^2 + v_3^2}}, \qquad r_2 = \frac{1}{2} \frac{(v_1 - \frac{1}{2})^2 + v_2^2 + v_3^2 + v_4^2}{\sqrt{v_1^2 + v_2^2 + v_3^2}}.$$

By the relations:

$$\frac{d}{dt} = \frac{1}{D} \frac{d}{ds}$$

$$\frac{d^2}{dt^2} = \frac{1}{D^2} \frac{d^2}{ds^2} - \frac{1}{D^3} \frac{dD}{ds} \frac{d}{ds},$$

the equations of motion in terms of the new time become:

$$Dx'' - D'x' - 2D^2y' = D^3\Omega_x$$
$$Dy'' - D'y' + 2D^2x' = D^3\Omega_y \qquad (15)$$
$$Dz'' - D'z' = D^3\Omega_z.$$

Notice that also in this case the singularities are eliminated. In fact, the singularities appear at the r.h.s. of (15), where $\Omega_x, \Omega_y, \Omega_z$ contain terms proportional to $\dfrac{1}{r_1^3}$ and $\dfrac{1}{r_2^3}$. Since D^3 is proportional to $r_1^3 r_2^3$, the quantities $D^3\Omega_x, D^3\Omega_y, D^3\Omega_z$ do not contain singularities at the attracting centers.

The only singularity still present corresponds to $v_1 = v_2 = v_3 = 0$, however this is not a true singularity, since these conditions correspond to $\bar{P}_3$ placed at ∞, which does not have physical meaning in our problem.

4   Numerical integration

We now show results obtained by numerical integration of the equations of motion relative to the planar circular three body problem before the Birkhoff regularization and after regularization. We recall that the Hamiltonian function describing the problem is given by:

$$H_{pre}(y_1, y_2, \tilde{p}_1, \tilde{p}_2) = \frac{1}{2}(\tilde{p}_1{}^2 + \tilde{p}_2{}^2) + y_2\tilde{p}_1 - y_1\tilde{p}_2 - \frac{1-\mu}{r_1} - \frac{\mu}{r_2}, \qquad (16)$$

while after the regularization, the Hamiltonian function assumes the form:

$$H_2(Q_1, Q_2, P_1, P_2) = \frac{1}{2D}[P_1{}^2 + P_2{}^2 + P_1\frac{\partial}{\partial Q_2}(f^2 + g^2)$$

$$-P_2\frac{\partial}{\partial Q_1}(f^2 + g^2) - (1 - 2\mu)(P_1 g_{Q_1} + P_2 f_{Q_1})] - (1 - \mu)\frac{2\rho}{\rho_1{}^2} - \mu\frac{2\rho}{\rho_2{}^2}. \qquad (19)$$

Following the numerical exploration of the restricted three body problem performed by Michel Hénon ([2]) we initially select a set of initial conditions for the positions, impulses and masses, and we integrate the equations of motion associated to the Hamiltonian (16) and (19). Then we plot the orbits on the plane (y_1, y_2), (Q_1, Q_2) respectively, and we compare the orbits corresponding to the same initial conditions before and after the regularization. More precisely, we fix a value of the Jacobi integral C relative to the Hamiltonian (16), and the initial conditions on the mass μ, on the position and on $\tilde{p}_1$, while we derive the initial condition on $\tilde{p}_2$ as a function of the energy integral. Then, by means of the relations which link the pre–regularized and post–regularized variables, we obtain the initial conditions relative to Q_1, Q_2, P_1, P_2, and we integrate the two systems of equations of motion for the Hamiltonian (16) and (19). Typically we have three kinds of dynamics: regular, chaotic and escape orbits. In

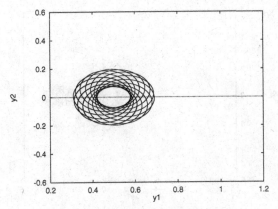

Fig. 1. Regular orbit pre–regularization

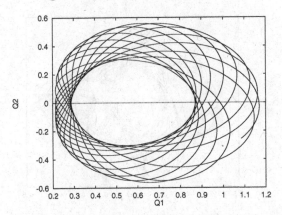

Fig. 2. Regular orbit post–regularization

Figs. 1, 2 we report a regular orbit in the (y_1, y_2)–plane and in the (Q_1, Q_2)–plane respectively, while Figs. 3, 4 show a chaotic orbit in the (y_1, y_2)–plane and in the (Q_1, Q_2)–plane respectively. We observe that after the regularization the primaries are placed at the positions $(-\frac{1}{2}, 0), (\frac{1}{2}, 0)$ respectively. Moreover, the orbits after the regularization are more wide, namely during close encounters the distances between the satellite and the primaries are larger after regularization. We do not show figures concerning escape orbits because they become unbounded in a very short time and it is not so interesting to show their graph.

The Hamiltonian formalism used to describe the Birkhoff transformation is a very useful tool for further studies concerning the application of perturbation theories to the Hamiltonian of the restricted three body problem, or for a numerical investigation of the behaviour of chaoticity indicators (Lyapunov exponents, local Lyapunov indicators, fast Lyapunov indicators) before and after the regularization.

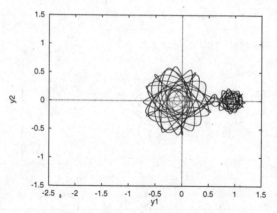

Fig. 3. Chaotic orbit pre–regularization

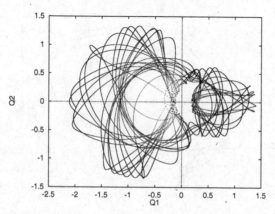

Fig. 4. Chaotic orbit post–regularization

References

1. G.D. Birkhoff: *Collected mathematical papers*, vol. **1**, Dover publications, INC., New York, (1968)
2. M. Henon: *Exploration numérique du problème restreint*, **I, II, III** parties, Annales d' Astrophysique, **28**, 1965.
3. P. Kustaanheimo, E. Stiefel: J. Reine Angew. Math **218**, 204 (1965)
4. E.L. Stiefel, M. Rössler, J. Waldvogel, C.A. Burdet: NASA Contractor Report **CR-769**, Washington, 1967
5. E.L. Stiefel, G. Scheifele: *Linear and Regular Celestial Mechanics*. (Springer–Verlag, Berlin, Heidelberg, New York 1971)
6. E.L. Stiefel, J. Waldvogel: C.R. Acad. Sci. Paris **260**, 805 (1965)
7. V. Szebehely: *Theory of orbits*. (Academic Press, New York and London 1967)
8. J. Waldvogel: Bull. Astron. **3**, II, 2, 295 (1967)

Perturbative Methods in Regularization Theory

Corrado Falcolini

Università di Roma Tre, Dipartimento di Matematica, L.go S.L. Murialdo 1, I-00146
Roma, Italy, e-mail: falco@mat.uniroma3.it

Abstract. We review the regularization procedure for a perturbed two-body problem
where the rescaled time variable is taken to be either the fictitious time or the generalized eccentric anomaly. We use Hamiltonian formalism to simplify the problem with
suitable canonical transformation and we make a comparison between the corresponding transformed systems. Using perturbative methods we also discuss the possibility of
analytical integration of the equations of motion and some related problem.

1 Introduction

In this lecture we want to make a comparison between two possible regularizations: one uses the original fictitious time variable (as in [1]) and the other the
generalized eccentric anomaly; the comparison is made on a perturbed two body
system using Hamiltonian formalism. Following [2], [3], [4], and especially [5],
we perform all transformations needed to put the Hamiltonian, and then the
equations of motion, in the best form to be handled with perturbative methods
and numerical integration procedures. In particular we use Kustaanheimo-Stiefel
transformation and Hamilton–Jacobi method. Hamiltonian formalism is shown
to be best suited for this since every transformation is found in canonical form
so that Hamiltonian structure is preserved and the analytical expansion that are
needed can be performed on a single function.

We present here only the two transformed Hamiltonians and the differences in
the related equations, but the choice of the independent time–like variable has
some effects also on the rate of convergence of the series involved and then also
on the accuracy in the results of integration: for instance the generalized eccentric anomaly variable is much more effective for very elongated orbits and even
in presence of collisions (see [5]).

Since perturbative methods strongly depend on the particular perturbation, we
give only an idea of how it is possible to get rigorous results using Fourier series
expansion and which are the problems that arise with this method.

2 Regularization procedure

In this chapter, we shortly review the general Kustaanheimo-Stiefel (K–S) regularization procedure and apply it to the fictitious time and generalized eccentric
anomaly cases. We also look for canonical transformations which could simplify
the two Hamiltonians.

2.1 The general case

Let us consider a generic time–dependent Hamiltonian of a three degrees of freedom system, that is a function of seven variables, with p conjugated to x

$$H(t, x, p) \qquad x, p \in \mathbb{R}^3 .$$

As we have already seen – in the K–S regularization – to regularize both the equations and the solutions one first has to consider an extended phase–space by letting $x_0 = t$ and $p_0 = -H$ along the solutions, with the resulting eight variables Hamiltonian

$$H_h(x_0, x, p_0, p) \equiv H(x_0, x, p) + p_0$$

then one has to pass to a rescaled time by introducing a new independent variable – "e.g." s – related to t in order to regularize solutions – "e.g." $dt/ds = D$ – to get a new Hamiltonian

$$\hat{H}(x_0(s), x(s), p_0(s), p(s)) \equiv DH_h$$

and finally one has to perform a coordinate change of variables to regularize also the equations of motion and the resulting Hamiltonian is forced to depend on two more variables

$$\tilde{H}(u_0, u, w_0, w) \qquad u, w \in \mathbb{R}^4 .$$

2.2 The fictitious-time case

Let us consider as an application, from know on, the time–dependent Hamiltonian of a perturbed two body problem as, for example, in a third body attraction both in the interior and in the external case:

$$H(t, x, p) = \frac{1}{2}|p|^2 - \frac{k^2}{r} + \varepsilon V(t, x) \qquad x, p \in \mathbb{R}^3 . \tag{1}$$

The corresponding homogeneous Hamiltonian H_h obtained, as we have said, by letting $x_0 = t$ and $p_0 = -H$ along the solutions is

$$H_h(x_0, x, p_0, p) = \frac{1}{2}|p|^2 - \frac{k^2}{r} + \varepsilon V(x_0, x) + p_0 \tag{2}$$

and passing to the rescaled (or "fictitious") time s by the relation

$$\frac{dt}{ds} = r$$

we get

$$\hat{H}_s(x_0(s), x(s), p_0(s), p(s)) = \frac{1}{2}|p|^2 r - k^2 + \varepsilon r V(x_0, x) + r p_0 .$$

In order to apply the K–S transformation we remind the basic formulae derived from the transformation matrix Λ, which is the 4×3 matrix

$$\Lambda(\boldsymbol{u}) \equiv \begin{pmatrix} u_1 & -u_2 & -u_3 & u_4 \\ u_2 & u_1 & -u_4 & -u_3 \\ u_3 & u_4 & u_1 & u_2 \end{pmatrix} ,$$

that is the change of variables, with $\boldsymbol{u}, \boldsymbol{w} \in \mathbb{R}^4$,

$$\boldsymbol{x} = \Lambda(\boldsymbol{u})\boldsymbol{u} \quad , \quad x_0 = u_0 \quad \Rightarrow \quad r \equiv |\boldsymbol{x}| = |\boldsymbol{u}|^2$$

$$\boldsymbol{p} = \frac{1}{2r}\Lambda(\boldsymbol{u})\boldsymbol{w} \quad , \quad p_0 = w_0 \quad \Rightarrow \quad |\boldsymbol{p}|^2 r = \frac{1}{4}|\boldsymbol{w}|^2$$

leading to the Hamiltonian

$$\tilde{H}_s(u_0, \boldsymbol{u}, w_0, \boldsymbol{w}) = \frac{1}{8}|\boldsymbol{w}|^2 - k^2 + \varepsilon|\boldsymbol{u}|^2 V(u_0, \boldsymbol{u}) + |\boldsymbol{u}|^2 w_0 .$$

It is finally useful to get the Hamiltonian in a slightly different form rescaling $\tilde{H}$ by $1/4$ and setting $u_0 = 2\bar{u}_0$, $w_0 = 2\bar{w}_0$, $\boldsymbol{u} = \bar{\boldsymbol{u}}$ and $\boldsymbol{w} = 4\bar{\boldsymbol{w}}$:

$$\bar{H}_s(\bar{u}_0, \bar{\boldsymbol{u}}, \bar{w}_0, \bar{\boldsymbol{w}}) = \frac{1}{2}|\bar{\boldsymbol{w}}|^2 - \frac{k^2}{4} + \frac{\varepsilon}{4}|\bar{\boldsymbol{u}}|^2 V(\bar{u}_0, \bar{\boldsymbol{u}}) + \frac{1}{2}\bar{w}_0|\bar{\boldsymbol{u}}|^2 . \tag{3}$$

Hamilton equations are then in closed form, given the initial conditions, for the Hamiltonian $\bar{H}_s$

$$\frac{d\bar{u}_0}{ds} = \frac{\partial \bar{H}_s}{\partial \bar{w}_0} = \frac{1}{2}|\bar{\boldsymbol{u}}|^2 = \frac{1}{2}r \quad \left(\text{``i.e.''} \ \frac{dt}{ds} = r\right)$$

$$\frac{d\bar{\boldsymbol{u}}}{ds} = \frac{\partial \bar{H}_s}{\partial \bar{\boldsymbol{w}}} = \bar{\boldsymbol{w}}$$

$$\frac{d\bar{w}_0}{ds} = -\frac{\partial \bar{H}_s}{\partial \bar{u}_0} = -\frac{\varepsilon}{4}|\bar{\boldsymbol{u}}|^2 \frac{\partial V}{\partial \bar{u}_0} \quad \left(\text{``i.e.''} \ \frac{d\bar{H}}{dt} = -\varepsilon\frac{\partial V}{\partial t}\right)$$

$$\frac{d\bar{\boldsymbol{w}}}{ds} = -\frac{\partial \bar{H}_s}{\partial \bar{\boldsymbol{u}}} = -\bar{w}_0\bar{\boldsymbol{u}} - \frac{\varepsilon}{4}\frac{\partial}{\partial \bar{\boldsymbol{u}}}|\bar{\boldsymbol{u}}|^2 V$$

Notice that these regularized equations contain both time transformation and energy law, that they are as simple as a perturbed armonic oscillator with frequency $\sqrt{\bar{w}_0}$ and that the frequency is constant if V is independent of t.

2.3 The generalized eccentric anomaly case

Let H be the Hamiltonian in (1) which we assume to be negative (elliptic case) and H_h as in (2) with $p_0 > 0$. A different regularization is obtained using, in place of s, the independent variable E – the generalized eccentric anomaly – defined by

$$\frac{dt}{dE} = \frac{r}{\sqrt{2p_0}}$$

which puts the Hamiltonian $\hat{H}_E \equiv \hat{H}_E(x_0(E), \boldsymbol{x}(E), p_0(E), \boldsymbol{p}(E))$ in the form

$$\hat{H}_E = \frac{1}{2}|\boldsymbol{p}|^2 \frac{r}{\sqrt{2p_0}} - \frac{k^2}{\sqrt{2p_0}} + \varepsilon \frac{r}{\sqrt{2p_0}} V(x_0, \boldsymbol{x}) + \frac{r}{2}\sqrt{2p_0} \ .$$

The K–S transformation gives in this case

$$\tilde{H}_E(u_0, \boldsymbol{u}, w_0, \boldsymbol{w}) = \frac{1}{8}\frac{|\boldsymbol{w}|^2}{\sqrt{2w_0}} - \frac{k^2}{\sqrt{2w_0}} + \varepsilon \frac{|\boldsymbol{u}|^2}{\sqrt{2w_0}} V(u_0, \boldsymbol{u}) + \frac{|\boldsymbol{u}|^2}{2}\sqrt{2w_0} \ .$$

It is possible to eliminate the dependence on the square root of w_0, keeping the advantages of the Hamiltonian formalism, by looking for a canonical tranformation given by a suitable generating function. It can be easily checked that the function

$$S(\bar{w}_0, \bar{\boldsymbol{w}}, u_0, \boldsymbol{u}) = \frac{2k}{\sqrt{-\bar{w}_0}} \sum_{i=1}^{4} u_i \bar{w}_i + \frac{k^4}{2} \frac{u_0}{\bar{w}_0^2}$$

generates the transformation

$$\bar{u}_0 = \frac{\partial S}{\partial \bar{w}_0} = \frac{-k^4}{\bar{w}_0^3} u_0 - \frac{1}{2\bar{w}_0} \sum_{i=1}^{4} \bar{u}_i \bar{w}_i$$

$$\bar{\boldsymbol{u}} = \frac{\partial S}{\partial \bar{\boldsymbol{w}}} = \frac{2k}{\sqrt{-\bar{w}_0}} \boldsymbol{u}$$

$$w_0 = \frac{\partial S}{\partial u_0} = \frac{k^4}{2\bar{w}_0^2}$$

$$\boldsymbol{w} = \frac{\partial S}{\partial \boldsymbol{u}} = \frac{2k}{\sqrt{-\bar{w}_0}} \bar{\boldsymbol{w}}$$

leading to the Hamiltonian

$$\bar{H}_E(\bar{u}_0, \bar{\boldsymbol{u}}, \bar{w}_0, \bar{\boldsymbol{w}}) = \frac{1}{2}|\bar{\boldsymbol{w}}|^2 + \bar{w}_0 + \varepsilon \frac{\bar{w}_0^2}{4k^4}|\bar{\boldsymbol{u}}|^2 V(\bar{u}_0, \bar{\boldsymbol{u}}, \bar{w}_0, \bar{\boldsymbol{w}}) + \frac{1}{8}|\bar{\boldsymbol{u}}|^2 \ . \qquad (4)$$

Hamilton equations for the Hamiltonian $\bar{H}_E$ are, also in this case and for any given initial conditions, in closed form

$$\frac{d\bar{u}_0}{dE} = \frac{\partial \bar{H}_E}{\partial \bar{w}_0} = 1 + \frac{\varepsilon}{4k^4}|\bar{\boldsymbol{u}}|^2 \frac{\partial}{\partial \bar{w}_0}\bar{w}_0^2 V$$

$$\frac{d\bar{\boldsymbol{u}}}{dE} = \frac{\partial \bar{H}_E}{\partial \bar{\boldsymbol{w}}} = \bar{\boldsymbol{w}} + \frac{\varepsilon}{4k^4}\bar{w}_0^2|\bar{\boldsymbol{u}}|^2 \frac{\partial V}{\partial \bar{\boldsymbol{w}}}$$

$$\frac{d\bar{w}_0}{dE} = -\frac{\partial \bar{H}_E}{\partial \bar{u}_0} = -\frac{\varepsilon}{4k^4}\bar{w}_0^2|\bar{\boldsymbol{u}}|^2 \frac{\partial V}{\partial \bar{u}_0}$$

$$\frac{d\bar{\boldsymbol{w}}}{dE} = -\frac{\partial \bar{H}_E}{\partial \bar{\boldsymbol{u}}} = -\frac{\bar{\boldsymbol{u}}}{4} - \frac{\varepsilon}{4k^4}\bar{w}_0^2 \frac{\partial}{\partial \bar{\boldsymbol{u}}}|\bar{\boldsymbol{u}}|^2 V \ .$$

Notice that regularized equations in this case are as simple as in the fictitious-time case but with some differences: 1) the frequency is always constant (equal to 1/2), 2) in the unperturbed case ($\varepsilon = 0$) besides $\bar{w}_0 = $ const one gets $\bar{u}_0 = E + $ const, 3) in the non conservative case the perturbation is more complicated since V depends also on $\bar{w}_0$ and $\bar{w}$.

3   Analytic perturbative methods

In this section we show how the two systems with Hamiltonian (3) and (4) can be analytically integrated, at least at first order, using Hamilton–Jacobi canonical transformations and we discuss integration at higher orders.

3.1   Hamilton–Jacobi for the fictitious-time case

First we look for a canonical transformation that would transform the Hamiltonian (3) in the unperturbed case, that is

$$H_s^0 = \frac{1}{2}|\bar{\boldsymbol{w}}|^2 + \frac{1}{2}\bar{w}_0|\bar{\boldsymbol{u}}|^2 = \frac{1}{2}\sum_{i=1}^{4}\bar{w}_i^2 + \frac{1}{2}\bar{w}_0\sum_{i=1}^{4}\bar{u}_i^2$$

in a new Hamiltonian which would depend only on the "action" variables. Let

$$S(\alpha_0, \boldsymbol{\alpha}, \bar{u}_0, \bar{\boldsymbol{u}}) = \alpha_0\bar{u}_0 + \sum_{i=1}^{4}\int\sqrt{2\alpha_i - \alpha_0\bar{u}_i^2}\,d\bar{u}_i$$

be the generating function which defines the transformation

$$\bar{u}_0 = \beta_0 + \frac{1}{2}\alpha_0^{-\frac{3}{2}}\sum_{i=1}^{4}\alpha_i[\sqrt{\alpha_0}\beta_i - \frac{1}{2}\sin(2\sqrt{\alpha_0}\beta_i)]$$

$$\bar{u}_i = \sqrt{\frac{2\alpha_i}{\alpha_0}}\sin(\sqrt{\alpha_0}\beta_i)$$

$$\bar{w}_0 = \alpha_0$$

$$\bar{w}_i = \sqrt{2\alpha_i}\cos(\sqrt{\alpha_0}\beta_i)\,,$$

leading to the transformed Hamiltonian

$$H_s^0(\beta_0(s), \boldsymbol{\beta}(s), \alpha_0(s), \boldsymbol{\alpha}(s)) = \alpha_1 + \alpha_2 + \alpha_3 + \alpha_4 \tag{5}$$

which obviously implies that β_0, α_0, $\boldsymbol{\alpha}$ are constants and $\boldsymbol{\beta}$ is linear with respect to s. If we now insert the solution of the unperturbed system with Hamiltonian (5) in the perturbed Hamiltonian (3) we get

$$H_s = \alpha_1 + \alpha_2 + \alpha_3 + \alpha_4 + \frac{\varepsilon}{2\alpha_0}\sum_{i=1}^{4}\alpha_i\sin^2(\sqrt{\alpha_0}\beta_i)V(\beta_0, \boldsymbol{\beta}, \alpha_0, \boldsymbol{\alpha})\,.$$

Finally, it is possible to eliminate the dependence on two variables in the trigonometric functions (which complicate the successive series expansion) making use again of the Hamiltonian formalism and looking for another canonical transformation such that $\bar{\beta}_i = \sqrt{\alpha_0}\beta_i$: the generating function

$$S(\alpha_0, \boldsymbol{\alpha}, \bar{\beta}_0, \bar{\boldsymbol{\beta}}) = -\alpha_0\bar{\beta}_0 + \frac{1}{\sqrt{\alpha_0}} \sum_{i=1}^{4} \alpha_i\bar{\beta}_i$$

defines the transformation

$$\bar{\alpha} = \frac{1}{\sqrt{\alpha_0}}\alpha \qquad \bar{\alpha}_0 = \alpha_0$$

$$\beta = \frac{1}{\sqrt{\alpha_0}}\bar{\beta} \qquad \beta_0 = -\frac{1}{2}\alpha_0^{-\frac{3}{2}} \sum_{i=1}^{4} \alpha_i\bar{\beta}_i + \bar{\beta}_0$$

and the final Hamiltonian

$$H_s^1 = \sqrt{\bar{\alpha}_0}(\bar{\alpha}_1 + \bar{\alpha}_2 + \bar{\alpha}_3 + \bar{\alpha}_4) + \frac{\varepsilon}{2\sqrt{\bar{\alpha}_0}} \sum_{i=1}^{4} \bar{\alpha}_i \sin^2 \bar{\beta}_i \, V(\bar{\beta}_0, \bar{\boldsymbol{\beta}}, \bar{\alpha}_0, \bar{\boldsymbol{\alpha}}) \,. \qquad (6)$$

The equations of motion are in the unperturbed case:

$$\bar{\boldsymbol{\alpha}} = \text{const} \qquad \bar{\alpha}_0 = \text{const}$$

$$\bar{\beta}_i = \sqrt{\bar{\alpha}_0}s + \text{const} \qquad \bar{\beta}_0 = \frac{1}{2\sqrt{\bar{\alpha}_0}}(\bar{\alpha}_1 + \bar{\alpha}_2 + \bar{\alpha}_3 + \bar{\alpha}_4)s + \text{const} \,.$$

3.2   Hamilton–Jacobi for the generalized eccentric anomaly case

The Hamiltonian (4) in the unperturbed case is now

$$H_E^0 = \frac{1}{2}|\bar{\boldsymbol{w}}|^2 + \frac{1}{8}|\bar{\boldsymbol{u}}|^2 + \bar{w}_0$$

and the generating function is, separating the additive element $\bar{w}_0$, that of an harmonic oscillator with frequency $1/2$

$$S(\alpha_0, \boldsymbol{\alpha}, \bar{u}_0, \bar{\boldsymbol{u}}) = \alpha_0\bar{u}_0 + \sum_{i=1}^{4} \int \sqrt{2\alpha_i - \frac{1}{4}\bar{u}_i^2} \, d\bar{u}_i \,.$$

The generated transformation

$$\bar{u}_0 = \beta_0$$

$$\bar{u}_i = 2\sqrt{2\alpha_i} \sin(\frac{\beta_i}{2})$$

$$\bar{w}_0 = \alpha_0$$

$$\bar{w}_i = \sqrt{2\alpha_i} \cos(\frac{\beta_i}{2})$$

inserted in the perturbed case, gives the new Hamiltonian

$$H_E = -\frac{k^2}{\sqrt{2\alpha_0}} + \alpha_1 + \alpha_2 + \alpha_3 + \alpha_4 + \frac{\varepsilon}{\alpha_0} \sum_{i=1}^{4} \alpha_i \sin^2(\frac{\beta_i}{2}) V(\beta_0, \boldsymbol{\beta}, \alpha_0, \boldsymbol{\alpha}) .$$

The solution of the equations of motion are, in the unperturbed case:

$$\boldsymbol{\alpha} = \text{const} \qquad \alpha_0 = \text{const}$$
$$\beta_i = E + \text{const} \qquad \beta_0 = k^2(2\alpha_0)^{-\frac{3}{2}} E + \text{const} .$$

3.3 Series expansion

Let us consider, as an example, the problem of a third body attraction with the perturbing body on a fixed periodic orbit of period $2\pi/\nu$. Using Fourier expansion we get

$$V = V(x_1, x_2, x_3, \cos(\nu t), \sin(\nu t), \ldots)$$

and using K–S transformation in the fictitious time case the perturbation becomes

$$V = V(u_1, u_2, u_3, u_4, \cos(2\nu u_0), \sin(2\nu u_0), \ldots) .$$

The Hamilton–Jacobi transformation gives simply

$$u_i = \sqrt{2\alpha_i}(\alpha_0)^{-\frac{1}{4}} \sin \beta_i$$

$$u_0 = \beta_0 - \frac{1}{4\alpha_0} \sum_{i=1}^{4} \alpha_i \sin(2\beta_i)$$

and setting $U = (\varepsilon/4)|\boldsymbol{u}|^2 V$ the Hamiltonian (6) can be written as

$$H_s^1 = \sqrt{\alpha_0}(\alpha_1 + \alpha_2 + \alpha_3 + \alpha_4) + U(\beta_0, \boldsymbol{\beta}, \alpha_0, \boldsymbol{\alpha}) .$$

If we look at the periods of the functions involved we see that the variables β_i are of constant period so that the u_i's are of period 2π in β_i, u_0 is of period π in β_i, $\cos(2\nu u_0), \sin(2\nu u_0)$ are of period π in β_i and π/ν in β_0 and finally U is of period 2π in β_i and π/ν in β_0. Notice that in the conservative case (for example in the oblateness problem) if V were a polynomial in the variables x_1, x_2, x_3, r, then U might be written as a sum of a finite number of Fourier components in β_i which is not the case, for instance, using Delaunay elements.

In the general case one can expand U using Fourier series in $\beta_0, \beta_1, \beta_2, \beta_3, \beta_4$, but the term $\cos(2\nu u_0)$ brings in an infinite number of components:

$$\cos(2\nu u_0) = \cos 2\nu(\frac{1}{4\alpha_0} \sum_{i=1}^{4} \alpha_i \sin(2\beta_i) - \beta_0)$$

and using the formula

$$\cos(A \sin \lambda + B) = \sum_{j=-\infty}^{+\infty} J_j(A) \cos(j\lambda + B)$$

where J_j are Bessel functions, one gets

$$\cos(2\nu u_0) = \sum_{j_1, j_2, j_3, j_4} \prod_{i=1}^{4} J_{j_i}\left(\frac{\nu\alpha_i}{2\alpha_0}\right) \cos 2(j_1\beta_1 + j_2\beta_2 + j_3\beta_3 + j_4\beta_4 - \nu\beta_0) .$$

So the potential U can be expanded in a fivefold Fourier series with respect to $\beta_0, \beta_1, \ldots, \beta_4$:

$$U = \varepsilon f_0(\alpha_0, \boldsymbol{\alpha}) + \varepsilon \sum_{n_0, \ldots, n_4} f_{n_0 \boldsymbol{n}}(\alpha_0, \boldsymbol{\alpha}) \exp[i(2\nu n_0\beta_0 + n_1\beta_1 + \cdots + n_4\beta_4)] .$$

Notice that one main advantage of Hamiltonian formalism is that this expansion is needed only for one function (H_s).

As usual in perturbation theory, it is often possible to push the perturbation to higher orders and to get an approximate solution by means of canonical transformations (i.e. $H = H_0 + \varepsilon V \to H' = H_0' + \varepsilon^2 V'$) to the expence of getting a much more complicated perturbing function.

3.4 First-order perturbation

As an example of rigorous qualitative results of analytic perturbative methods, we derive first order equations of the previous example.

Let $\boldsymbol{n} \cdot \boldsymbol{\beta} \equiv n_1\beta_1 + \cdots + n_4\beta_4$ and the Hamiltonian be

$$H_s^1 = \sqrt{\alpha_0}(\alpha_1 + \alpha_2 + \alpha_3 + \alpha_4) + \varepsilon f_0(\alpha_0, \boldsymbol{\alpha})$$
$$+ \varepsilon \sum_{n_0, \ldots, n_4} f_{n_0 \boldsymbol{n}}(\alpha_0, \boldsymbol{\alpha}) \exp[i(2\nu n_0\beta_0 + \boldsymbol{n} \cdot \boldsymbol{\beta})] ,$$

then the corresponding Hamilton equations are

$$\frac{d\alpha_i}{ds} = -\varepsilon \sum_{n_0 \boldsymbol{n}} n_i f_{n_0 \boldsymbol{n}}(\alpha_0, \boldsymbol{\alpha}) \exp[i(2\nu n_0\beta_0 + \boldsymbol{n} \cdot \boldsymbol{\beta})]$$

$$\frac{d\alpha_0}{ds} = -2\varepsilon\nu \sum_{n_0 \boldsymbol{n}} n_0 f_{n_0 \boldsymbol{n}}(\alpha_0, \boldsymbol{\alpha}) \exp[i(2\nu n_0\beta_0 + \boldsymbol{n} \cdot \boldsymbol{\beta})] \qquad (7)$$

$$\frac{d\beta_i}{ds} = \sqrt{\alpha_0} + \varepsilon \frac{\partial f_0}{\partial \alpha_i}(\alpha_0, \boldsymbol{\alpha}) + \varepsilon \sum_{n_0 \boldsymbol{n}} \frac{\partial f_{n_0 \boldsymbol{n}}}{\partial \alpha_i}(\alpha_0, \boldsymbol{\alpha}) \exp[i(2\nu n_0\beta_0 + \boldsymbol{n} \cdot \boldsymbol{\beta})]$$

$$\frac{d\beta_0}{ds} = \frac{1}{2\sqrt{\alpha_0}}(\alpha_1 + \alpha_2 + \alpha_3 + \alpha_4) + \varepsilon \frac{\partial f_0}{\partial \alpha_0}(\alpha_0, \boldsymbol{\alpha})$$
$$+ \varepsilon \sum_{n_0 \boldsymbol{n}} \frac{\partial f_{n_0 \boldsymbol{n}}}{\partial \alpha_0}(\alpha_0, \boldsymbol{\alpha}) \exp[i(2\nu n_0\beta_0 + \boldsymbol{n} \cdot \boldsymbol{\beta})] .$$

Since the variables $\alpha_0, \alpha_1, \ldots, \alpha_4$ are constant in the unperturbed motion, the solution

$$\alpha_0 = \alpha_0^0 = \text{const} \qquad \alpha_i = \alpha_i^0 = \text{const}$$
$$\beta_0 = \beta_0^0 + \frac{1}{2\sqrt{\alpha_0^0}}(\alpha_1^0 + \alpha_2^0 + \alpha_3^0 + \alpha_4^0) \qquad \beta_i = \beta_i^0 + \sqrt{\alpha_0^0}s$$

can be plugged in the system (7) after the expansion of the terms $\sqrt{\alpha_0}$ and $1/\sqrt{\alpha_0}$.

The system contains now sums of trigonometric terms with arguments:

$$2\nu n_0 \beta_0^0 + n_1 \beta_1^0 + \cdots + n_4 \beta_4^0 + [\frac{\nu n_0}{\sqrt{\alpha_0^0}}(\alpha_1^0 + \cdots + \alpha_4^0) + \sqrt{\alpha_0^0}(n_1 + \cdots + n_4)]s .$$

The integration of such terms produces a secular perturbation for all $n_0, \boldsymbol{n}$ such that

$$\frac{\nu n_0}{\sqrt{\alpha_0^0}}(\alpha_1^0 + \alpha_2^0 + \alpha_3^0 + \alpha_4^0) + \sqrt{\alpha_0^0}(n_1 + n_2 + n_3 + n_4) = 0 ,$$

i.e. in presence of isotropic resonances:

$$n_0 = 0 \quad \text{and} \quad n_1 + n_2 + n_3 + n_4 = 0$$

or physical resonances:

$$n_0 \neq 0 \quad \text{and} \quad 2\nu(\alpha_1^0 + \alpha_2^0 + \alpha_3^0 + \alpha_4^0)n_0 + \alpha_0^0(n_1 + n_2 + n_3 + n_4) = 0$$

It is then possible to see that the resonance conditions depend on the mean motion ν of the perturbing body and on the initial values $\alpha_0^0, \alpha_1^0, \ldots, \alpha_4^0$ of the particle.

References

1. P. Kustaanheimo, E. Stiefel: J. Reine Angew. Math **218**, 204 (1965)
2. G. Janin, V.R. Bond: Adv. Space Res. **1**, 69 (1981)
3. U. Kirchgraber: Celestial Mechanics **8**, 251 (1973)
4. E.L. Stiefel, M. Rössler, J. Waldvogel, C.A. Burdet: NASA Contractor Report **CR-769**, Washington, 1967
5. E.L. Stiefel, G. Scheifele: *Linear and Regular Celestial Mechanics*. (Springer–Verlag, Berlin, Heidelberg, New York 1971)

Collisions and Singularities in the n-body Problem

Corrado Falcolini

Università di Roma Tre, Dipartimento di Matematica, L.go S.L. Murialdo 1, I-00146 Roma, Italy, e-mail: falco@mat.uniroma3.it

Abstract. We present recent results on the existence of non-collision singularities and on the regularizability of multiple or simultaneous collisions in the solution of the n-body problem. The first problem is obviously only theoretical but its solution gave new impulse to recent studies on dynamical systems applied to space travels and planetary sciences. We review the main results which led to Xia's and Gerver's examples which for the first time gave an affirmative proof of Painlevé's conjecture. The second problem is only sketched, listing several papers and results of the last ten years, and has been one of the many ingredients in Xia's proof.

1 Introduction

The long lasting conjecture made by Painlevé more than a century ago has been clarified and proved in the last thirty years. We present here a short history of the conjecture and the main participants to its solution. We also describe the problem and the solutions obtained in recent years as well as related problems and new questions which remain still unanswered. We have mainly used the references [4], [9], [11], [24], [31] for the past and present history of the problem and the original articles for the presentation of its different solutions.

We would like to cite a remark of J. Moser presenting the proof of Painlevé's conjecture as one of the significant advances in the field of dynamical systems during the last 50 years ([24]): "Clearly this solution is not of any astronomical significance. Why do I present it: It shows, in one example, the progress gained from the study of hyperbolical dynamical systems which provided the understanding and the tools for the solution of this problem. It also reminds us of the efforts that go in the studies of singularities in partial differential equations, e.g. of the Navier–Stokes equation, provided they exist! One usually thinks of singularities as a local phenomenon, but even this (simple!) classical example of ordinary differential equations exhibits such complicated singularities of nonlocal type, whose existence was doubted for a long time".

2 Non-collision singularities in Newtonian systems

In this chapter, we shortly review the history of the problem starting from Painlevé and von Zeipel.

2.1   The n-body problem

Given n points P_i with masses m_i, positions $q_i \in \mathbb{R}^3$ and velocities $\dot{q}_i \in \mathbb{R}^3$ we have

$$m_i \ddot{q}_i = \sum_{j \neq i} \frac{m_i m_j}{r_{ij}^3}(q_i - q_j) = \frac{\partial U}{\partial q_i} \qquad (i = 1, ... n) \qquad (1)$$

where $r_{ij} \equiv |q_i - q_j|$ is the distance between P_i and P_j; U is the potential

$$U = \sum_{j < i} \frac{m_i m_j}{r_{ij}}$$

which is real analytic on $\mathbb{R}^{3n} \setminus \Delta$ where Δ is the "collision set"

$$\Delta \equiv \bigcup_{i<j} \Delta_{ij} \equiv \bigcup_{i<j} \{q \equiv (q_1, ..., q_n) \in \mathbb{R}^{3n} \text{ s.t. } q_i = q_j\}$$

the set of configurations where $r_{ij} = 0$ for some $i \neq j$ (i.e. collisions) and where a "singularity" of the equation of motion (1) occurs.

Here we would like to use the word *singularity* for the solution of (1):

Definition 1: A solution $q(t)$ of (1) has a *singularity* at time $t^* < \infty$ if it cannot be analitically extended beyond t^*.

Obviously (in the n-body problem) a collision is also a singularity. Is any singularity always a collision? The question was only partially answered by Painlevé and it is still known as "Painlevé's conjecture" ([25]).

Theorem 1 (Painlevé, 1895): The n-body problem has the property that $q(t)$ has a singularity at $t = t^*$ if and only if $q(t) \to \Delta$ as $t \to t^*$.

But the condition $q(t) \to \Delta$ does not necessarily imply a collision:

Definition 2: A singularity at time t^* is a *collision singularity* if there exists $\delta \in \Delta$ such that $q(t) \to \delta$ as $t \to t^*$.

Definition 3: A *non-collision singularity* is a singularity which is not a collision singularity. (That is $\lim_{t \to t^*} q(t)$ would not exist on Δ).

Do non-collision singularities exist?

Theorem 2 (Painlevé): For $n = 3$ all singularities are collision singularities.

Painlevé's conjecture: For $n \geq 4$ the n-body problem admits solutions with non-collision singularities.

Let

$$I = \sum_{i=1}^{n} \frac{1}{2} m_i |q_i|^2$$

be the moment of inertia of the system (which is also a measure of the size of the system). It is easy to see that the limit of I as $t \to t^*$ always exists: in fact the singularity at t^* implies that $q(t) \to \Delta$ as $t \to t^*$ which means that the potential U and (by Lagrange's Identity) also $\ddot{I} = U + 2h$ tends to infinity so that, near t^*, $\ddot{I}(t)$ is greater than 0 which is a sufficient condition for the existence of the limit.

In 1908, in a four pages article communicated by A. Lindstedt to a Swedish journal ([43]), H. von Zeipel gave the first main contribution to the problem raised by Painlevé. The theorem that now takes his name had alternate consideration among the scientific community: it was reformulated independently by Chazy in 1920 ([3]), it was considered incomplete by Wintner in his 1941 book on celestial mechanics ([41]), it was then fixed in details by Sperling ([36]) thirty years later to be finally recognized as correct by McGehee in 1986 ([22]) couriously only two years before Xia's thesis with the first proof on the existence of non-collision singularities.

Theorem 3 (von Zeipel, 1908): If a non-collision singularity occurs at t^* then $\lim_{t \to t^*} I(t) = \infty$.

In other words, a non-collision singularity can occur only if the system, or some of its parts, becomes *unbounded in finite time*.

How can a particle go to infinity in finite time? In a different problem we can make a trivial example: the scalar differential equation $\dot{x} = x^2$ with initial condition $x(0) = x_0 > 0$ has the explicit solution $x(t) = x_0/(1 - x_0 t)$ which tends to infinity as $t \to t^* = 1/x_0$. In the n–body problem such a particle would still have to acquire an infinite amount of kinetic energy which is not forbidden since U is not bounded from below: what is needed is a very close encounter with another particle not only once but infinitely many times. How can this be accomplished for more than three particles?

The idea to overcome the problem is related to *double* and *triple collisions*: they were studied already by Sundman ([37]) in 1912, Siegel ([32]) in 1941 and Levi-Civita ([17]), but their work dealt with collision orbits with no description of the flow near these singularities. Starting from numerical experiments on Pythagorean problem in the late 60's ([38]) and then explained with triple encounters in the 70's ([21], [40]) it was made clear that the result of going close to a triple collision can be the formation of a binary system with the third particle moving far away with enormous speed. In the same years, research on the theoretical aspects about collisions singularities in the n-body problem continued with important results due mainly to Siegel and Moser ([33]), Pollard and Saari ([27], [28], [29]) , Conley and Easton ([5], [6], [12]).

In his 1974 paper ([21]), originated by a collaboration with Conley, Mc Gehee used "polar coordinates" to blow up the singularity set, for a triple collision in the collinear three-body problem, and to replace it with an invariant boundary called *collision manifold*, extending the dynamical system to a new flow on a augmented phase space which is much easier to study on the boundary. This new idea was going to be the key tool in Xia's proof.

Collision singularities correspond to hyperbolic rest points and orbits that reach collision in finite time approach these rest points as the new rescaled time ("blow-up") approaches infinity. The problem is then reduced to study the stable and unstable manifolds of these rest points.

3   The solution

3.1   The example of Mather and McGehee

The first breakthrough towards the solution of the problem was made in 1975 by Mather and McGehee ([20]) for the collinear four-body problem: four point masses (m_1, m_2, m_3, m_4) on a stright line with m_1 equal to m_2 forming a binary system which moves towards minus infinity, m_4 greater than m_1 leading to plus infinity and m_3 much smaller than m_1 going back and forth with increasing velocity. It was the first example of an unbounded solution in finite time in a newtonian problem but this result was achieved with an infinite number of regularized binary collisions and therefore it did not solve Painlevé's Conjecture.

With a correct timing argument (that is with an appropriate *symbolic dinamic* proof) double collisions (m_3, m_4) and close to triple collisions (m_1, m_2, m_3) can repeat infinitely often within a finite period of time. The authors were able to prove that there exists, for this particular system, a Cantor set of initial conditions leading to such a particular behaviour.

3.2   The first example of Gerver

Gerver's tentative solution appeared in 1984, ([15]), and dealt with a planar five-body problem: m_1, m_3, m_4 of comparable size, m_2 much larger than m_1 and m_5 much smaller than the other masses. The methods used by Gerver were "elementary", did not required new technique, but he needed a huge amount of calculations for a complete proof of the correct behaviour of his model.

He imagined m_1 to be on an almost circular orbit around m_2 with m_3 and m_4 at a much larger distance and m_2, m_3, m_4 at the vertices of a triangle which is moving homotopically. m_5 moves rapidly around the other four bodies with $|\dot{q}_5| >> |\dot{q}_1|$.

The revolution time of m_5 around the triangle decreases each time it passes close to m_1 and m_2, since it picks up some kinetic energy from the binary system which reduce the distance between m_1 and m_2 but is enough to compensate a faster expansion of the triangle due to the interaction of m_5 with the masses at its vertices.

This example was only heuristic: it didn't work, or at least Gerver was not able to prove it in all details. What he presented had an explicit title: "A *possible* model for a singularity without collisions..." recognizing his impossibility to find a complete proof of such a five-body model.

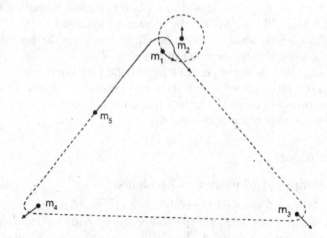

Fig. 1. Gerver's first example

3.3   The example of Xia

The first complete proof of Painlevé's conjecture was finally obtained in Xia's thesis in 1988 (later adjusted and fixed with his 1992 article in the Annals of Mathematics). Xia, ([42]), considered two binaries (m_1, m_2) and (m_3, m_4) on opposite planes perpendicular to the z-axis with m_1, m_2 and m_3, m_4 symmetric with respect to the z-axis where m_5 is moving. Such isosceles configurations, expecially in connection with triple collisions, where studied extensively by Simó, Devaney and Moeckel (see [34], [8] and [23]. A new approach on simmetric properties in the n-body problem can be found in [1] which had a featured review from Math. Rev. in 1998). In [34], [8] a large open set of "allowable" masses is found, which means that if properly organized these masses behave like in [21], [40] with one particle (here m_5) acquiring enormous speed. It is interesting to note that this behaviour is possible with m_5 significantly heavier than two equal masses forming a binary and that this is precisely the case that works also in Xia's example.

The proof for this spatial five-body model is stated using McGehee coordinates together with a lot of previous results on the subject and contains also many new ideas; it also generalize to any number of particles greater than five. In this model the center of mass is fixed at the origin, accounting for only six degree of freedom; setting the angular momentum to zero (the two binaries rotating in opposite directions) and fixing the energy one can consider a ten-dimensional phase space Ω.

Let $q(x; t)$ be the trajectory of the system with initial condition x and $\Sigma \subset \Omega$ be the set of initial conditions which end up in a triple collision of P_1, P_2, P_5 with central configuration. The result is stated in the following Theorems:

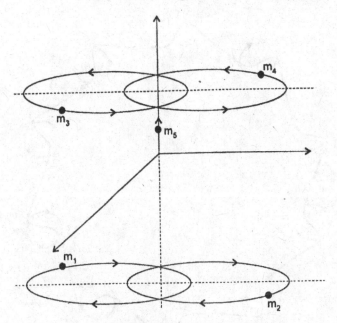

Fig. 2. Xia's example

Theorem 4 (Xia, 1992): There exist $m_1 = m_2$, $m_3 = m_4$, m_5 so that one can find $x^* \in \Sigma$ such that $q_4(x^*; t^*) = q_5(x^*; t^*)$, "i.e." x^* ends up in a simultaneous triple (P_1, P_2, P_5) and binary (P_3, P_4) collision. Moreover, for some 3-dimensional surface $\Pi \subset \Omega$ with $x^* \in \Pi \cap \Sigma$, there exist an uncountable set $\Lambda \subset \Pi$ with the property that for any $x \in \Lambda$ there exists $0 < t^* < \infty$ such that $q(x; t)$ is defined on $[0, t^*)$ (possibly with binary collisions) and satisfies $z_1(t) = z_2(t) \to -\infty$, $z_3(t) = z_4(t) \to +\infty$ for $t \to t^*$.

In other words there exist unbounded solutions in finite time (possibly with binary collisions).

Theorem 5 (Xia, 1992): Let $x^* \in \Sigma$, $\Lambda \subset \Pi \subset \Omega$ as in Theorem 4. Then one can find $x^* \in \Sigma$ so that there exists $\Lambda_0 \subset \Lambda$ such that for any $x \in \Lambda_0$ (also uncountable) $q_1(x; t) \neq q_2(x; t)$, $q_3(x; t) \neq q_4(x; t)$ for any $t \in [0, t^*)$.

In other words the solutions starting from Λ_0 experience a non-collision singularity.

3.4   The second example of Gerver

Gerver's second solution appeared in 1991 ([16]) and dealt with a planar $3n$-body problem (n large): n pairs of equal masses $m_{i_1} = m_{i_2}$ with $i = 1, \ldots, n$ on a regular polygon and n particles of smaller mass m_{i_3} moving around it. When a small particle interacts closely with a binary it gains part of the kinetic energy

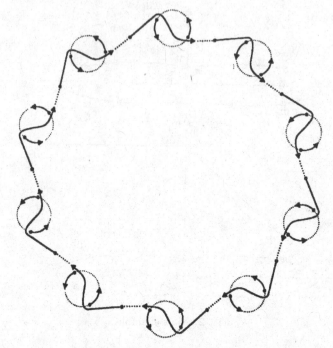

Fig. 3. Gerver's second example

of the pair and exchange with the binary part of its momentum, forcing the two particles of the binary to come closer to their baricenter, whose distance from the center of the polygon increases. The idea is the same of his previous attempt, but he made two improvements: the use of rotational symmetry to simplify the problem and consequently the introduction of a new free parameter, n, to get more freedom in the hard part of its calculations. The first planar example of Painlevé's conjecture can then be stated in the following way: there exists an integer n, suitable values of m_{i_1}, m_{i_3} $(i = 1, \ldots, n)$ and of initial velocities such that the system becomes unbounded in finite time avoiding all possible collisions.

4   Multiple and simultaneous binary collisions

About multiple collisions I will only refer to [26], [2], [10], [18], [39], [7] as some of the latest articles on a subject which has been extensively studied in many previous works and, as we have seen, has been a crucial tool on both Xia's and Gerver's construction.

Also simultaneous binary collisions (SBC) entered in Xia's proof and even if their nature was known to be very close to that of a single binary collision (they are algebraic branch points, [36], and the motion nearby closely mimic the behavior of a single binary collision) several results have been obtained recently.

Simó and Lacomba ([35]) have proved that SBC for the n-body problem and in any dimension are C^0-block regularizable in the sense of Easton's C^k-block regularization: near a SBC orbit there exists a C^k diffeomorphism connecting collision and near collision orbits with ejection and near ejection orbits and the motion can be continued beyond the SBC maintaining continuity with respect to initial conditions (see [12]); their result has been used by Xia and cited in a preprint form (see [42]).

El Bialy ([13], [14]) has shown that SBC in one dimension are C^1-block regularizable and that the series expansion of the SBC singularity has coefficients which depends analytically on SBC initial conditions.

Martinez and Simó ([19]) have used a geometric approach to get more insight into the problem and numerical evidence that the degree of differentiability, in the planar four-body SBC problem, of the block regularization is exactly 8/3.

5 Open questions

There are several questions which remain still open (see [31]):

- Is Painlevé's conjecture true for $n = 4$?
- Are there planar examples with n small ?
- Are there mass choices for which non-collision singularities cannot occur?
- Let C_n, NC_n, S_n be the sets of initial conditions for the n-body problem leading to a solution of (1) with, respectively, a collision singularity, a non-collision singularity and a singularity of any kind. We know that the Lebesgue measure of C_n, $\mu(C_n)$, is equal to zero for any n and also that $\mu(NC_4) = 0$ ([30]) so that the set of initial conditions leading to a singularity $\mu(S_n) = 0$ for $n \leq 4$. The same is true for those non-collision orbits where the particles eventually line up along a line (as in Xia's example but not in Gerver's). Is it true that $\mu(S_n) = 0$ for $n \geq 5$?
- If $\mathcal{CO}$ represents the set of initial conditions leading to a collision of any kind, is it true that $\overline{\mathcal{CO}}$ represents the set of initial conditions causing any kind of singulariy ?

The last question has been answered affirmatively by Saari and Xia ([31]) without a proof: as usual in Mathematics, solving a long lasting conjecture set the basis for new and interesting questions. We hope that Saari–Xia's conjecture will not need another century to be eventually proved.

References

1. A. Albouy, A. Chenciner: Invent. Math. **131**, 151 (1998)
2. M. Alvarez, J. Llibre: Extracta Math. **13**, 73 (1998)
3. J. Chazy: C.R. Hebdomadaires Seances Acad. Sci. Paris **170**, 575 (1920)
4. A. Chenciner: Asterisque **245**, 323 (1997)

5. C. Conley: 'Twist mappings, linking, analiticity and periodic solutions which pass close to an unstable periodic solution'. In: *Topological Dynamics* (Benjamin, New York 1968) pp.129–153

6. C. Conley, R.W. Easton: Trans. Amer. Math. Soc. **158**, 35 (1971)

7. J.M. Cors, J. Llibre: Nonlinearity **9**, 1299 (1996)

8. R. Devaney: Invent. Math. **60**, 249 (1980)

9. F.N. Diacu: Math. Intell. **15**, 6 (1993)

10. F.N. Diacu: J. Diff. Eq. **128**, 58 (1996)

11. F.N. Diacu, P. Holmes: *Celestial Encounters* (Princeton University Press, Princeton NJ 1996)

12. R.W. Easton: J. Diff. Eq. **12**, 361 (1972)

13. M.S. ElBialy: J. Diff. Eq. **102**, 209 (1993)

14. M.S. ElBialy: J. Math. Anal. Appl. **203**, 55 (1996)

15. J.L. Gerver: J. Diff. Eq. **52**, 76 (1984)

16. J.L. Gerver: J. Diff. Eq. **89**, 1 (1991)

17. T. Levi-Civita: *Sur la résolution qualitative du problème restreint des trois corps* (Opere Matematiche **2**, Bologna 1956)

18. Z. Liu, S. Zhang: J. Math. Anal. Appl. **212**, 343 (1997)

19. R. Martinez, C. Simó: Nonlinearity **12**, 903 (1999)

20. J. Mather, R. McGehee: Lecture Notes in Physics **38**, 573 (1975)

21. R. McGehee: Invent. Math. **27**, 191 (1974)

22. R. McGehee: Expo. Math. **4**, 335 (1986)

23. R. Moeckel: American J. Math. **103**, 1323 (1981)

24. J. Moser: Documenta Mathematica (Extra Volume ICM 1998) **1**, 381 (1999)

25. P. Painlevé: *Leçons sur la Théorie Analitique des Equations Différentielles* (A. Hermann, Paris 1897)

26. E. Pérez-Chavela, L.V. Vela-Arévalo: J. Diff. Eq. **148**, 186 (1998)

27. H. Pollard, D.G. Saari: Arch. Rational Mech. Anal. **30**, 263 (1968)

28. H. Pollard, D.G. Saari: 'Singularities of the n-body problem'. In: *Inequalities, II*, Proc. Second Sympos., U.S. Air Force Acad., Colo., 1967 (Academic Press, New York 1970) pp.255–259

29. D.G. Saari: Arch. Rational Mech. Anal. **49**, 311 (1973)

30. D.G. Saari: J. Diff. Eq. **26**, 80 (1977)

31. D.G. Saari, Z. Xia: Notices of AMS **42**, 538 (1995)

32. C.L. Siegel, Ann. of Math. (2) **42**, 127 (1941)

33. C.L. Siegel, J. Moser: *Lectures on Celestial Mechanics* (Springer–Verlag, Berlin 1971)

34. C. Simó: 'Analysis of triple collision in the isosceles problem'. In: *Classical Mechanics and Dynamical Systems* (Marcel Dekker, New York 1980)

35. C. Simó, E.A. Lacomba: J. Diff. Eq. **98**, 241 (1992)

36. H. Sperling: J. Reine Angew. Math. **245**, 15 (1970)

37. K. Sundman: Acta Math. **36**, 105 (1912)

38. V. Szebehely: Proc. Nat. Acad. Sci. USA **58**, 60 (1967)

39. H. Umeara, K. Tanikawa: 'Triple collision and escape in the three-body problem'. In: *Dynamical systems and chaos, Vol.2 (Hachioji, 1994)* (World Sci. Publishing, River Edge, NJ, 1995)

40. J. Waldvogel: Celestial Mech. **11**, 429 (1975)

41. A. Wintner: *The Analytical Foundations of Celestial Mechanics* (Princeton University Press, Princeton NJ 1941)

42. Z. Xia: Ann. Math. **135**, 411 (1992)

43. H. von Zeipel: Archiv. för Mat. Astron. och Fys. **4**, n. 32 (1908)

Triple Collision and Close Triple Encounters

Jörg Waldvogel

Applied Mathematics, ETH, 8092 Zurich, Switzerland

Abstract. In gravitational systems of point masses binary collisions are mathematically simple and well understood. Collisions of three or more particles are much more complicated, i.e. a dramatic increase of complexity occurs when the number N of particles involved in a collision increases from 2 to 3. Collisions of more than three particles seem to be of the same complexity as triple collisions. However, there are still unanswered questions concerning general N-body collisions.

The reason for the complexity of triple collision is the inherent sensitivity to initial conditions for solutions passing near triple collision, even after a short time. Specifically, a solution passing near triple collision may change dramatically if the initial conditions prior to the close encounter are modified infinitesimally. In contrast, this is not the case for a binary collision.

We use the planar three-body problem as a model in order to discuss the main features of triple collision of point masses and of its realistic counterpart, the close triple encounter. This comparatively simple model allows us to study all important aspects of close encounters of $N > 2$ gravitationally interacting point masses.

In Chapters 1 and 2 we discuss classical results, beginning with the equations of motion, then studying relationships between the total angular momentum and triple collision. C. L. Siegel's famous series for triple collision solutions, one time considered the highlight of the theory of triple collision, conclude the traditional part of these lectures.

Chapter 3 is devoted to studying the relationship between solutions engaging in a sharp triple collision and neighbouring solutions. The variational equation gives a rough idea of what is happening. A complete understanding can be achieved by means of R. McGehee's concept of the collision manifold, which arises by introducing special coordinates blowing up all possible states close to triple collision. In this context, possibilities of regularizing triple collision will be considered.

The close triple encounter may be seen as a particle accelerator that allows to accelerate point masses to arbitrarily large velocities. This is the key to the existence of the long-sought non-collision singularities in the motion of a sufficient number of gravitationally interacting point masses.

1 Basics

A convenient model for studying collision singularities in celestial mechanics is the gravitational N-body problem, i.e. the problem of the motion of N gravitationally interacting point masses in $\mathbb{R}^3$. For triple collisions to be possible $N \geq 3$ is necessary. It turns out that the three-body problem, $N = 3$, shows all the essential features of triple collision.

1.1   The general three-body problem: equations of motion, integrals of motion

We begin by defining the notation and collecting basic relationships [15,31]. Unless otherwise specified, these relations hold for systems of $N > 1$ point masses. Particular properties of the case $N = 3$ will be considered in the following sections.

Let $m_j > 0$, $j = 1, \ldots, N$ be N point masses located at the barycentric positions $x_j \in \mathbb{R}^3$, i.e.,

$$\sum_{j=1}^{N} m_j x_j = 0. \tag{1}$$

With the force function

$$U(x) := \sum_{1 \le j < k \le N} \frac{m_j m_k}{|x_j - x_k|}, \tag{2}$$

where $x = (x_1; x_2; \ldots; x_N) \in \mathbb{R}^{3N}$, the equations of motion are

$$m_j \ddot{x}_j = \frac{\partial}{\partial x_j} U(x) =: U_{x_j}(x), \quad j = 1, \ldots, N, \tag{3}$$

where dots denote derivatives with respect to time t.

By introducing the momenta

$$p_j := m_j \dot{x}_j, \quad j = 1, \ldots, N \tag{4}$$

satisfying $\sum_{j=1}^{N} p_j = 0$ the kinetic energy becomes

$$T := \frac{1}{2} \sum_{j=1}^{N} m_j |\dot{x}_j|^2 = \frac{1}{2} \sum_{j=1}^{N} \frac{|p_j|^2}{m_j}, \tag{5}$$

and the energy integral states that the total energy (or the Hamiltonian) H is constant on an orbit:

$$H := T - U = h = \text{const.} \tag{6}$$

Furthermore, the angular momentum C is a constant of motion:

$$C = \sum_{j=1}^{N} m_j x_j * \dot{x}_j = \sum_{j=1}^{N} x_j * p_j = \text{const.}, \tag{7}$$

where $*$ denotes the vector product.

Homogeneity. Some of the basic properties of collisions of all particles in an N-body problem are intimately connected with the homogeneity of the force function U and the kinetic energy T. We therefore state Euler's classical theorem on homogeneous functions:

Theorem 1. *Let $F : x = (x_1, \ldots, x_n) \in \mathbb{R}^n \longmapsto F(x) \in \mathbb{R}$ be homogeneous of degree μ, i.e. for every $\varepsilon > 0$, $x \in \mathbb{R}^n$ we have $F(\varepsilon x) = \varepsilon^\mu F(x)$. Then*

$$\sum_{k=1}^{n} x_k \frac{\partial F(x)}{\partial x_k} = \mu \cdot F(x). \tag{8}$$

Proof. Differentiate the homogeneity relation (8) with respect to ε and put $\varepsilon = 1$.
$\perp$

According to the definition (2), $U(x)$ is homogeneous of degree -1, and its gradient is homogeneous of degree -2:

$$U(\varepsilon x) = \varepsilon^{-1} U(x), \quad U_{x_j}(\varepsilon x) = \varepsilon^{-2} U_{x_j}(x). \tag{9}$$

On introducing new scaled variables $\widetilde{x}, \widetilde{t}$ according to

$$x = \varepsilon \widetilde{x}, \quad t = \varepsilon^{\frac{3}{2}} \widetilde{t} \quad (\varepsilon > 0) \tag{10}$$

into the equations of motion (3) we immediately deduce

$$p = \varepsilon^{-\frac{1}{2}} \widetilde{p}, \quad H = \varepsilon^{-1} \widetilde{H}, \quad C = \varepsilon^{\frac{1}{2}} \widetilde{C}, \tag{11}$$

where $\widetilde{p}, \widetilde{H}, \widetilde{C}$ are the scaled momenta, energy, and angular momentum. Inserting this into (3) yields

Theorem 2. [23]: *The equations of motion (3) of the N-body problem are invariant under the scaling transformation (10):*

$$m_j \frac{d^2}{d\widetilde{t}^2} \widetilde{x}_j = \frac{\partial}{\partial \widetilde{x}_j} U(\widetilde{x}), \quad j = 1, \ldots, N.$$

1.2   Angular momentum, Sundman's theory

The definitions, relations and theorems of this section are fundamental for understanding triple collision [15,31].

(a) **Sundman's inequality** (N-bodies)

We define the (polar) moment of inertia I as

$$I := \sum_{j=1}^{N} m_j |x_j|^2. \tag{12}$$

Differentiating twice and using (5), (6) yields Lagrange's formula,

$$\frac{1}{2} \ddot{I} = 2T - U \quad (= T + h = U + 2h), \tag{13}$$

which is the basis for **Sundman's inequality**:

$$2IT \geq \frac{1}{4}\dot{I}^2 + \frac{1}{N}|C|^2, \tag{14}$$

see, e.g. [15], p. 25. Integration of this differential inequality yields

(b) Sundman's theorem

Theorem 3. *If at time $t = t_1$ all N bodies of an N-body system in $\mathbb{R}^3$ collide, then the total angular momentum C vanishes for all times $t < t_1$.*

Remarks. (i) This implies that in an N-body collision the velocity vector of every particle has a limiting direction. A gradual spiraling in of all bodies is excluded.

(ii) The contraposition of Theorem 3 states: If an N body problem has the total angular momentum $C \neq 0$, the collision of all N bodies is excluded.

(c) Sundman's main result

Whereas the previous results hold for any $N > 1$, we now restrict ourselves to the three-body problem, $N = 3$. At the beginning of the 20th century this problem was considered one of the major unsolved problems of mathematics, and the "solution" most people had in mind was to express the motion of the three bodies in terms of known functions. Sundman's result [18] has to be seen in this context.

Consider a three-body problem with $C \neq 0$. Due to Remark (ii) above triple collision is excluded: therefore the only possible singularities are binary collisions. These may be regularized by means of the **fictitious time**

$$s := \int_0^t (1 + U(x(\tau))\,d\tau \tag{15}$$

(see the second chapter by A. Celletti and the chapter by G. Della Penna, in this book, as well as [7,20]). Sundman showed that the motion of the three bodies is described by functions which are analytic in the strip

$$D := \{s \in \mathbb{C}, \; |Im(s)| < \delta, \; \delta > 0\}$$

of the complex s-plane. Since D is mapped onto the unit disk $|z| < 1$ by the conformal map

$$s \longmapsto z = \tanh\left(\frac{\pi}{4}\frac{s}{\delta}\right) \tag{16}$$

the motion may be represented by Taylor series convergent for all positive and negative times.

Since, in a way, these series constitute the solution of the three-body problem, K.F. Sundman was awarded the prize the French Academy of Sciences had issued for solving the problem of three bodies. However, beautiful as Sundman's result may be, it is of little practical use since the series converge extremely slowly and suffer from catastrophic cancellation for large $|t|$.

(d) The three-body problem in $\mathbb{R}^3$ with $C = 0$

Theorem 3 states that $C = 0$ is a necessary condition for a triple collision in the three-body problem. $C = 0$, however, is not sufficient. On the other hand, we have

Theorem 4. *The three-body problem with $C = 0$ is planar, i.e. the motion takes place in a fixed plane.*

Proof. Assume that at time $t = t_0$ all position vectors x_j and velocity vectors $p_j (j = 1, 2, 3)$ are coplanar. Use a rectangular coordinate system with the third axis perpendicular to this plane, such that $x_j^{(3)}(t_0) = p_j^{(3)}(t_0) = 0$, $(j = 1, 2, 3)$. The equations of motion (3) imply $x_j^{(3)}(t) = p_j^{(3)}(t) = 0$ $(j = 1, 2, 3)$ for all times t. $\perp$

There remains to be shown that $C = 0$ at time t_0 implies coplanarity of the initial vectors $x_j = x_j(t_0)$, $p_j = p_j(t_0)$, $(j = 1, 2, 3)$.

Lemma 1. *Let $m_j > 0$, and $x_j \in \mathbb{R}^3$, $p_j \in \mathbb{R}^3$, $(j = 1, 2, 3)$ be column vectors such that*

$$\sum_{j=1}^{3} m_j x_j = 0, \quad \sum_{j=1}^{3} p_j = 0, \quad C := \sum_{j=1}^{3} x_j * p_j = 0. \tag{17}$$

Then the vectors x_j, p_j are **coplanar**.

Proof. Choose x_1, x_2, $p_1 \in \mathbb{R}^3$. Using (17_1) and (17_2) the condition (17_3) reads

$$C = x_1 * p_1 + x_2 * p_2 + \frac{1}{m_3}(m_1 x_1 + m_2 x_2) * (p_1 + p_2) = 0.$$

With the abbreviations

$$u := x_1\left(1 + \frac{m_1}{m_3}\right) + x_2\frac{m_2}{m_3}, \quad v := x_1\, dis\frac{m_1}{m_3} + x_2\left(1 + \frac{m_2}{m_3}\right) \in \mathbb{R}^3 \tag{18}$$

this may be written as the following system of linear equations for determining the column vector p_2:

$$S(v)\, p_2 = -S(u)\, p_1. \tag{19}$$

Here $S(v)$ is the skew-symmetric matrix

$$S(v) := \begin{bmatrix} 0 & -v_3 & v_2 \\ v_3 & 0 & -v_1 \\ -v_2 & v_1 & 0 \end{bmatrix} = -(S(v))^T$$

converting the vector product $v * p$ into the matrix product $S(v)p$. $S(v)$ has the property $S(v)\, v = 0$; therefore $S(v)$ is a singular matrix, $\det(S(v)) = 0$. The necessary and sufficient condition for the existence of solutions p_2 to the singular system (19) of linear equations is obtained by multiplying (19) by v^T:

$$v^T S(u) p_1 = 0 \quad \text{or} \quad \det(v, u, p_1) = 0.$$

Therefore the vectors v, u, p_1 are coplanar. Together with (18) this implies coplanarity of the vectors p_1, x_1, x_2. Coplanarity of p_2, x_1, x_2 is shown analogously. $\perp$

2   Triple collision

Simple and explicit examples of three-body motion leading to triple collision are the so-called **homothetic** solutions. These, in turn, are intimately connected with the homogeneity properties discussed in Section 1.1.

2.1   Homographic and homothetic solutions

Definition. A solution of (3) of the form

$$x_j(t) = r(t) \cdot \Omega(t) \cdot \xi_j \in \mathbb{R}^3, \quad j = 1, \dots, N, \tag{20}$$

where $\Omega(t)$ is an orthogonal 3-by-3 matrix, $(\Omega(t))^T \Omega(t) = Id$, is called **homographic**. In particular, if $\Omega(t) = Id$, the solution (20) is called **homothetic**. $\dashv$

Substitution of (20) into the equations of motion (3) yields, by using (9),

$$m_j \frac{d^2}{dt^2} \left(r(t) \, \Omega(t) \right) \xi_j = r(t)^{-2} \, \Omega(t) \, U_{\xi_j}(\xi), \tag{21}$$

from where conditions on the rotation matrix $\Omega(t)$, the radial factor $r(t)$, and the configuration $\xi = (\xi_1; \xi_2; \dots; \xi_N)$ may be derived. There follows immediately that the configuration ξ must satisfy

$$U_{\xi_j} = -\mu \, m_j \, \xi_j \tag{22}$$

for every j with an appropriate factor μ independent of j. Configurations with this property are called **central configurations** (see Section 2.2).

The most subtle aspect is the determination of the rotation matrix $\Omega(t)$. Its behaviour is governed by the following rather deep theorem.

Theorem 5. *The matrix $\Omega(t)$ describes a rotation with a* **fixed** *axis.*

Proof. See, e.g., Wintner [31], §372 - §374 bis (p. 288 - 292). The length of the proof in this book known for its highly concentrated style illustrates the depth of Theorem 5. In fact, for non-Newtonian central forces homographic solutions can "tumble". At the end of the long proof Wintner writes (§374 bis):

"One might think the preceding proof is unnecessarily
complicated; in fact, it seems plausible that the theorem
is a direct consequence of the homogeneity of the force
function U and the conservation of angular momentum.
Such is, however, not the case."

$\perp$

Based on Theorem 5 the radial factor $r(t)$ follows easily from (21):

Corollary 1. *Let* $z(t) := r(t)\,\Omega(t)\,\zeta$, *where* $\zeta \in \mathbb{R}^3$ *is a fixed vector, then* $z(t)$ *satisfies the differential equation of Kepler motion,*

$$\ddot{z}(t) = -\mu\,\frac{z(t)}{|z(t)|^3}\,. \tag{23}$$

In summary, in homographic solutions the N particles move on synchronized Keplerian orbits in a fixed plane such that the configuration remains similar to itself. Figs. 1 and 2 show the two types of elliptic homographic motion of $N = 3$ bodies generated by the two possible central configurations: an equilateral triangle (Lagrange) and a collinear arrangement (Euler).

Homothetic solutions are particular homographic solutions with no rotation, i.e. the Kepler motion (23) is rectilinear. Therefore every homothetic solution of the N-body problem ($N \geq 2$) leads to simultaneous collisions of all N bodies.

2.2 Central configurations

In this section the algebraic problem of condition (22) will be discussed, roughly following [31], see also [19].

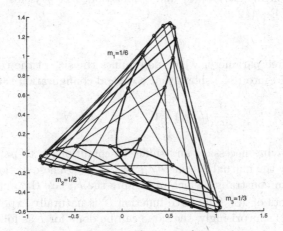

Fig. 1. Lagrangean homographic motion

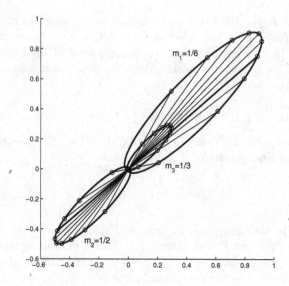

Fig. 2. Eulerian homographic motion

Definition. $\xi_j \in \mathbb{R}^d (j = 1, \ldots, N, d \geq 1)$ is called a **central configuration** with respect to the masses $m_j > 0$ if $\sum_{j=1}^{N} m_j \xi_j = 0$ and there exists a constant μ such that

$$U_{\xi_j} = -\mu\, m_j\, \xi_j, \quad j = 1, \ldots, N. \tag{24}$$

$\perp$

Multiplication of (24_j) by ξ_j and summation over j yields, together with Theorem 1 and Eq. (12):

$$\mu = \frac{U}{I}, \tag{25}$$

i.e. $\mu > 0$ is a free parameter which determines the size of the configuration ξ. Substituting (25) into (24) shows that a central configuration satisfies

$$\frac{\partial}{\partial \xi_j} (U^2 I) = 0, \quad j = 1, \ldots, N. \tag{26}$$

One way of deriving necessary and sufficient conditions is to parameterize the configuration by the mutual distances $r_{jk} := |\xi_j - \xi_k|$. These distances may have to satisfy certain constraints which make sure the r_{jk} are the mutual distances between N points of $\mathbb{R}^d$. The force function U is naturally expressed in terms of r_{jk} by Eq. (2). Surprisingly, the same can be done for I as follows. Consider

mI and use algebraic identities:

$$mI - 0 = (m_1 + m_2 + \ldots + m_N)(m_1 \xi_1^2 + \ldots + m_N \xi_N^2) -$$
$$(m_1 \xi_1 + \ldots + m_N \xi_N)^2$$

$$= \sum_{j<k} m_j m_k (\xi_j - \xi_k)^2 = \sum_{j<k} m_j m_k |\xi_j - \xi_k|^2 \,.$$

In view of this and Eq. (26) we have

Theorem 6. *Let $m_j > 0$, $(j = 1, \ldots, N)$ and*

$$U(\xi) = \sum_{1 \leq j < k \leq N} \frac{m_j m_k}{|\xi_j - \xi_k|} \,, \quad \xi = (\xi_1; \ldots; \xi_N) \in \mathbb{R}^{Nd}$$

$$I(\xi) = \frac{1}{m} \sum_{1 \leq j < k \leq N} m_j m_k |\xi_j - \xi_k|^2, \quad m = \sum_{j=1}^{N} m_j \,. \tag{27}$$

Necessary and sufficient for ξ to be a central configuration is

$$\delta(U^2 I) = 0 \,, \tag{28}$$

or $U^2 I$ is stationary under the geometric constraints making sure that

$$r_{jk} := |\xi_j - \xi_k|, \quad 1 \leq j < k \leq N$$

are the $N(N-1)/2$ mutual distances of N distinct points of $\mathbb{R}^d$.

Using the geometric constraints asks for the technique of Lagrange multipliers. An alternative is to normalize the size of the configuration and to describe it by parameters $\rho_1, \rho_2, \ldots, \rho_f$, where f is the number of degrees of freedom, such that the geometric constraints are automatically satisfied. Since condition (28) is invariant under non-degenerate coordinate transformations, it now becomes

$$\frac{\partial}{\partial \rho_k} (U^2 I) = 0, \quad k = 1, \ldots, f \,. \tag{29}$$

Examples. (i) No constraints, e.g., $N = 3$ points of $\mathbb{R}^2$, the 3 distances r_{12}, r_{23}, r_{13} may be chosen independently.

We have

$$U = m_1 m_2 r_{12}^{-1} + m_2 m_3 r_{23}^{-1} + m_1 m_3 r_{13}^{-1},$$
$$mI = m_1 m_2 r_{12}^2 + m_2 m_3 r_{23}^2 + m_1 m_3 r_{13}^2,$$

and (28) yields

$$\frac{\partial}{\partial r_{jk}} (U^2 \cdot mI) = U^2 \cdot 2 m_j m_k r_{jk} + 2U \left(-\frac{m_j m_k}{r_{jk}^2} \right) \cdot mI = 0 \,.$$

There follows

$$r_{jk}^3 = \frac{mI}{U}, \quad 1 \le j < k \le N, \tag{30}$$

i.e. all three mutual distances are equal. For $N = 3$ the central configuration is the Lagrangean configuration of an equilateral triangle (even for unequal masses!).

The same reasoning applies for $N = 2$ points of the line $\mathbb{R}^1$: the central configuration consists of two distinct points of R^1, the "one-dimensional equilateral simplex".

Analogously, we may have $N = 4$ points of $\mathbb{R}^3$ and 6 independent mutual distances. According to Eq. (30) the central configuration for any set of masses is the regular tetrahedron.

(ii) $N = 3$ points of $\mathbb{R}^1$, the collinear (Eulerian) configuration of three masses m_1, m_2, m_3. Three different central configurations are possible according as m_1, m_2, or m_3 is the inner mass. We consider the arrangement (m_1, m_2, m_3), normalize the configuration by $r_{23} = 1$ and introduce $\rho := r_{12} > 0$ as (the only) independent parameter. With $r_{13} = \rho + 1$ we obtain

$$mI = m_1 m_2 \rho^2 + m_1 m_3 (\rho + 1)^2 + m_2 m_3,$$
$$U = \frac{m_1 m_2}{\rho} + \frac{m_1 m_3}{\rho + 1} + m_2 m_3.$$

Eq. (29) now becomes $d(U^2 \cdot mI)/d\rho = 0$; it leads to the well-known 5th-degree equation for ρ, which may be written in matrix form as

$$(m_1 \; m_2 \; m_3) \begin{bmatrix} 0 & 0 & 0 & -3 & -3 & -1 \\ 1 & 2 & 1 & -1 & -2 & -1 \\ 1 & 3 & 3 & 0 & 0 & 0 \end{bmatrix} \begin{bmatrix} \rho^5 \\ \rho^4 \\ \rho^3 \\ \rho^2 \\ \rho \\ 1 \end{bmatrix} = 0. \tag{31}$$

It may be shown that for $m_j > 0$ Eq. (31) has exactly one real solution ρ with $0 < \rho < 1$. Clearly, the inner mass m_2 plays a distinguished role, whereas the interchange of m_1 and m_3 corresponds to replacing ρ with $1/\rho$, as may also be deduced from geometric considerations.

2.3   Triple collision, Siegel's series

In view of the remark at the end of Section 2.1 it is natural to look at phase space near homothetic solutions in order to describe triple collisions. We will concentrate on the problem of $N = 3$ bodies in the plane $\mathbb{R}^2$ (see Theorems 3 and 4). The basic plan is to use a simple homothetic solution as a reference and study nearby solutions by linearization techniques.

As our reference solution we choose

$$x_j^0(t) := c \cdot t^{\frac{2}{3}} \cdot \xi_j, \quad j = 1, 2, 3, \quad \xi_j \in \mathbb{R}^2, \tag{32}$$

where ξ is a Lagrangean or an Eulerian central configuration, and $c := (4.5\mu)^{\frac{1}{3}}$, [21]. A convenient choice is $\mu := m = m_1 + m_2 + m_3$ which normalizes the central configurations in a natural way. Eq. (32) corresponds to choosing $z(t)$ of Eq. (23) as the rectilinear parabolic Kepler motion.

In order to take advantage of the center-of-mass integrals we first rewrite the equations of motion (3) in terms of relative coordinates y_1, y_2 with respect to x_3:

$$y_1 := x_1 - x_3, \quad y_2 := x_2 - x_3, \quad y := \begin{pmatrix} y_1 \\ y_2 \end{pmatrix} \in \mathbb{R}^4.$$

The transformed equations of motion are of the form

$$\ddot{y} = f(y) \tag{33}$$

where

$$f(y) = - \begin{bmatrix} (m_1 + m_3)\, y_1\, r_1^{-3} + m_2\, y_2\, r_2^{-3} + m_2(y_1 - y_2)\, r_{12}^{-3} \\ (m_2 + m_3)\, y_2\, r_2^{-3} + m_1\, y_1\, r_1^{-3} + m_1(y_2 - y_1)\, r_{12}^{-3} \end{bmatrix} \tag{34}$$

with $r_1 := |y_1|$, $r_2 := |y_2|$, $r_{12} := |y_1 - y_2|$. The vector-valued function f is homogeneous of degree -2, i.e.

$$f(\varepsilon y) = \varepsilon^{-2} f(y). \tag{35}$$

Let therefore

$$y_0(t) := c\, t^{\frac{2}{3}} \begin{pmatrix} \xi_1 - \xi_3 \\ \xi_2 - \xi_3 \end{pmatrix} \tag{36}$$

be our reference solution; we investigate the family of nearly solutions $y(t) = y_0(t) + \eta(t)$, where $\eta(t)$ is a sufficiently small perturbation. In first order η satisfies the variational equation of (33) with respect to the reference y_0:

$$\ddot{\eta}(t) = t^{-2} J_0 \cdot \eta(t), \tag{37}$$

where $J_0 \in \mathbb{R}^{4 \times 4}$ is a constant matrix obtained by substituting the central configuration ξ_j into the Jacobi matrix of (34).

If J_0 is diagonalizable (37) may be solved in terms of power functions, using the 4 eigenvectors e_k and eigenvalues λ_k of J_0 satisfying

$$J_0\, e_k = e_k\, \lambda_k, \quad k = 1, 2, 3, 4. \tag{38}$$

We obtain the 8 particular solutions

$$\eta_k^{(1)}(t) = e_k \cdot t^{\mu_k^{(1)}}, \quad \eta_k^{(2)}(t) = e_k\, t^{\mu_k^{(2)}}, \quad k = 1, 2, 3, 4, \tag{39}$$

where $\mu_k^{(1)}$, $\mu_k^{(2)}$, ordered as $\mu_k^{(1)} > \mu_k^{(2)}$ if real, are the solutions of the quadratic equation

$$\mu_k^2 - \mu_k - \lambda_k = 0, \quad k = 1, 2, 3, 4. \tag{40}$$

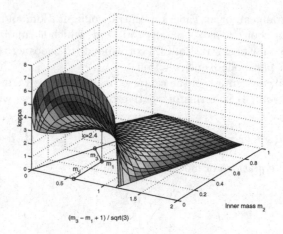

Fig. 3. The parameter κ versus the mass triangle in the Eulerian case

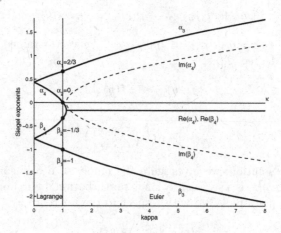

Fig. 4. The Siegel exponents versus κ

By defining Siegel's exponents α_k, β_k [13] as

$$\alpha_k := \mu_k^{(1)} - \frac{2}{3}, \quad \beta_k := \mu_k^{(2)} - \frac{2}{3}, \quad k = 1, 2, 3, 4 \tag{41}$$

the 8-parameter family of neighbouring solutions of $y_0(t)$ may be approximated by the linear family

$$y(t) = ct^{\frac{2}{3}} \left[e_1 + \sum_{k=1}^{4} e_k (a_k \, t^{\alpha_k} + b_k \, t^{\beta_k}) \right]. \tag{42}$$

Here $a_k \ll 1$, $b_k \ll 1$, $k = 1, 2, 3, 4$ are the parameters of the family. A time shift in (36) shows that the first eigensystem is given by $e_1 = (\xi_1 - \xi_3; \xi_2 - \xi_3)$, $\lambda_1 = \frac{4}{9}$, which implies $\alpha_1 = \frac{3}{2}$, $\beta_1 = -1$.

Solving the eigenvalue problem (38) and the quadratic equation (40) shows that 4 Siegel exponents (denoted by $\alpha_1, \alpha_2, \beta_1, \beta_2$) are independent of the masses. These constant exponents are associated with classical constants of motion:

$$\alpha_1 = \tfrac{2}{3}: \quad \text{conservation of energy}$$
$$\alpha_2 = \quad 0: \quad \text{rotation of the coordinate system}$$
$$\beta_2 = -\tfrac{1}{3}: \quad \text{conservation of angular momentum}$$
$$\beta_1 = -1: \quad \text{invariance with respect to time shift}$$

The 4 variable Siegel exponents $\alpha_3, \alpha_4, \beta_3, \beta_4$ are found to be - more or less - explicit functions of the masses [13,24]. First define the auxiliary parameter κ as follows:

$$\kappa := \begin{cases} \dfrac{1}{m}\sqrt{\dfrac{1}{2}\left((m_1 - m_2)^2 + (m_2 - m_3)^2 + (m_3 - m_1)^2\right)} \in [0,1], \\ \qquad\qquad\qquad\qquad\qquad\qquad\qquad \text{Lagrangean case}, \\[2mm] \dfrac{m(m_1\, u^{-3} + m_2\, u^{-3}\, v^{-3} + m_3\, v^{-3})}{(m_1 + m_2(u^{-2} + v^{-2}) + m_3)^2} \in [1,8], \\ \qquad\qquad\qquad\qquad\qquad\qquad\qquad \text{Eulerian case}. \end{cases} \tag{43}$$

In the Eulerian case $u := r_{12}/r_{13}$, $v := r_{23}/r_{13}$ are the ratios of the distances in the central configuration of the masses (m_1, m_2, m_3); the dependence of κ on the masses is shown in Fig. 3. Then, the variable Siegel exponents may be expressed by κ alone as

$$\left.\begin{array}{c}\alpha_k \\ \beta_k\end{array}\right\} = \frac{1}{6}\left(-1 \pm \sqrt{1 + \gamma(1 + \sigma\kappa)}\right),$$

where

$$\gamma = \begin{cases} 12, & \kappa \le 1 \\ 8, & \kappa > 1 \end{cases}, \qquad \sigma = \begin{cases} 4 - \tfrac{\gamma}{4}, & k = 3 \\ -1, & k = 4 \end{cases}.$$

Figure 4 visualizes the dependence of the variable exponents on κ as well as the fixed Siegel exponents.

Having discussed the range of values of the exponents α_k, β_k in the linear family (42), we return to the question of describing all solutions in the family that have a triple collision at $t = 0$. Clearly, terms with negative exponents must not be present, e.g. in the Lagrangean case $b_1 = b_2 = b_3 = b_4 = 0$. By suppressing the term $a_2 t^0$ in (42) by means of an appropriate rotation of the coordinate system, Siegel [13] proved

Theorem 7. *If none of the two exponents $\alpha_1, \alpha_3, \alpha_4$ are commensurable the solutions with a triple collision at $t = 0$ may be written as*

$$y(t) = c t^{\frac{2}{3}} P(a_1 t^{\frac{2}{3}}, a_3 t^{\alpha_3}, a_4 t^{\alpha_4}), \tag{44}$$

where P is a multiple power series convergent for sufficiently small $|t| < t_0$, and a_1, a_3, a_4 are free parameters.

Remarks. (i) Eq. (44) holds in the Lagrangean case; in the Eulerian case the last argument of P is absent.

(ii) If some of the exponents have a rational ratio, logarithmic terms occur, see [14]. E.g., if $\alpha_3 = n \cdot \frac{2}{3}$, $(n \in \mathbb{N})$ the second argument in (44) is $a_3 \, t^{\alpha_3} \log t$ instead of $a_3 \, t^{\alpha_3}$.

(iii) If we think of time running backwards, $t \searrow 0$, to produce a triple collision at $t = 0$, there can be no physically meaningful analytic continuation of the motion for $t < 0$ since (44) generally yields complex values for $t < 0$. Therefore, regularization in the sense of, e.g., Levi-Civita is not possible.

The question about the significance of the negative Siegel exponents (absent from (44)) arises naturally. The terms of (42) with negative real parts in the exponents become singular as $t \to 0$; however, they have meaningful limits (namely 0) as $t \to \infty$. Therefore it makes sense to consider the subfamily $a_1 = a_3 = a_4 = 0$ as $t \to \infty$. All three bodies escape parabolically as $t \to \infty$; the motion is referred to as triple-parabolic escape.

Theorem 8. [25]. *If none of the two exponents $\beta_2, \beta_3, \beta_4, \alpha_4$ are commensurable the triple-parabolic escape solutions may be written as*

$$y(t) = c(t - \tau)^{\frac{2}{3}} \, P(b_2 \, t^{-\frac{1}{3}}, \; b_3 \, t^{\beta_3}, \; b_4 \, t^{\beta_4}, \; a_4 \, t^{\alpha_4}), \qquad (45)$$

where P is a multiple power series convergent for sufficiently large $|t| > t_0$, and b_2, b_3, b_4, a_4 are free parameters. In the Lagrangean case the last argument of P is absent.

Remarks. (i) In (45) the power series argument $b_1 \, t^{-1}$ has been suppressed by shifting the time.

(ii) For $\kappa > 1.125$ (Eulerian case only) α_4 and β_4 are complex, $\beta_4 = \overline{\alpha}_4$. Since the corresponding terms of (42) add up to

$$t^{\frac{2}{3}} \left(a_4 \, t^{\alpha_4} + \overline{a}_4 \, t^{\overline{\alpha}_4} \right) = 2 \, |a_4| \, \sqrt{t} \, \cos\left(\omega \, log\frac{t}{t_0} \right), \quad \omega = Im(a_4),$$

the Eulerian triple-parabolic escape may show a slow oscillation growing as $O(\sqrt{t})$. However, this oscillation is only visible if the leading term of (42) vanishes, e.g. in symmetric cases, $m_1 = m_3$ [25].

The fact that continuation of an orbit beyond a sharp triple collision is generally not possible raises the question about the behaviour of three-body motion at a close triple encounter (not leading to a sharp collision of all bodies) and beyond.

3   The close triple encounter

For understanding close triple encounters the homogeneity relations (10), (11) play an essential role [22,23]. If no sharp triple collision occurs the motion undergoes at most binary-collision singularities. By regularization and continuation through the singularities the motion may be continued for all times. Therefore, there must be a minimum size of the system, denoted by $\varepsilon > 0$.

We distinguish two cases: (i) ε is known in advance as, e.g., in the case of a third body being "shot" into a binary of size ε. Then the scaling transformation (10), (11) generally transforms the motion into one with no immediate triple close encounter.

(ii) The three bodies approach their center of mass nearly in a central configuration. Then the three-body system may collapse to arbitrarily small size $\varepsilon > 0$, where ε is not known in advance. Therefore this case cannot be handled by predetermined scaling of the variables.

3.1   Singular perturbations

Under the assumption that the motion is a member of the family (42), a system of 8 linear equations will determine the values of the family parameters $a_k, b_k, k = 1, \ldots, 4$; assume all of them to be small of the order $O(\varepsilon)$.

In the following we will sketch a singular-perturbation approach [1] to describing three-body motion in case (ii), using three phases of the motion [26]. This discussion will lead to a heuristic understanding of close triple encounters; a rigorous treatment will be sketched in Section 3.2.

Assume therefore that y is a member of the family (42) not leading to a sharp triple collision, and that all family parameters are proportional to a single small parameter $\varepsilon > 0 : a_k = O(\varepsilon), \ b_k = O(\varepsilon), \ k = 1, \ldots, 4$. The parabolic collapse must break up at some scale $\delta = \delta(\varepsilon) > 0$. In the limit $\varepsilon \to 0$ we obviously must have $\delta \to 0$; it suffices to assume $\delta(\epsilon) = \epsilon^\sigma$ with an unknown positive exponent $\sigma > 0$.

Rewriting (42) in terms of scaled variables $\widetilde{y}, \widetilde{t}$, where, according to (10), (11),

$$y = \delta \widetilde{y}, \ t = \delta^{\frac{3}{2}} \widetilde{t}, \ p = \delta^{-\frac{1}{2}} \widetilde{p}, \ H = \delta^{-1} \widetilde{H}, \ C = \delta^{\frac{1}{2}} \widetilde{C}, \tag{46}$$

results in

$$\widetilde{y} = c\widetilde{t}^{\frac{2}{3}} \left[e_1 + \sum_{k=1}^{4} e_k (a_k \, \delta^{1.5\alpha_k} \, \widetilde{t}^{\alpha_k} + b_k \, delta^{1.5\beta_k} \, \widetilde{t}^{\beta_k}) \right]. \tag{47}$$

Due to $a_k = O(\varepsilon)$, $\delta^{1.5\alpha_k} = O(\varepsilon^{1.5\sigma\alpha_k})$ the limit $\varepsilon \to 0$ exists if

$$1 + \frac{3}{2}\sigma\alpha_k \geq 0 \ \text{ and } \ 1 + \frac{3}{2}\sigma\beta_k \geq 0 \ \text{ for } \ k = 1, \ldots, 4. \tag{48}$$

This implies

$$\sigma \leq \frac{-2/3}{\min\limits_{k}(Re\,\alpha_k, Re\,\beta_k)} = \frac{2/3}{-\beta_3} \ ;$$

hence the largest region for $\delta(\varepsilon)$ (the "distinguished limit" [1]) is given by $\sigma = 2/(-3\beta_3)$. In the limit $\varepsilon \to 0$ we then have, as follows from (47) with $\delta = \varepsilon^\sigma$,

$$\widetilde{y} = c\widetilde{t}^{\frac{2}{3}} \left[e_1 + e_3 \, \frac{b_3}{\varepsilon} \, \widetilde{t}^{\beta_3} + o(1) \right]. \tag{49}$$

For sufficiently large $\tilde{t}$ Eq. (49) constitutes initial conditions for the "inner solution". If $\tilde{t}$ varies in the full range, $-\infty < \tilde{t} < \infty$, the inner solution describes the triple close encounter in terms of the scaled variables $\tilde{t}, \tilde{y}$. This motion is the particular solution of the three-body problem under consideration given by the initial conditions (49), to be satisfied asymptotically as $\tilde{t} \to -\infty$. In view of (46_4) and (46_5) this motion satisfies $\tilde{H} = 0$, but generally $\tilde{C} \neq 0$.

The behaviour **after** the close encounter is determined by the final evolution of the inner solution as $\tilde{t} \to +\infty$. According to Chazy [2] the "allure finale" of zero-energy three-body motion ($\tilde{H} = 0$) can be hyperbolic-elliptic expansion or triple parabolic expansion. In the generic first case one of the bodies, m_e, $e \in (1, 2, 3)$ escapes with momentum $\tilde{p}_e < \infty$, whereas the other two bodies form a binary escaping to the opposite side. Transforming back according to (46_3) yields

$$p_e = \delta^{-\frac{1}{2}} \tilde{p}_e \,,$$

i.e. the momentum of the escaping body can be arbitrarily large, whereas the other two bodies form an arbitrarily tight binary after the triple close encounter.

The case of the inner solution ending in parabolic expansion as $\tilde{t} \to \infty$ has measure zero within the family (49). The corresponding triple close encounter orbit may be continued by an appropriate triple explosion, generally along a different central configuration. In contrast to regularization of binary collisions by analytic continuation this is an example of regularization by surgery [5,16].

Triple close encounters may be visualized in 8-dimensional phase space by means of invariant manifolds [27] (see Fig. 5). Due to $\alpha_2 = 0$ there is a 1-dimensional center manifold C^1 in the form of a circle corresponding to the rotational invariance of the problem. The incoming orbits or collision orbits (no exponents < 0) form the center-stable manifold S^d of dimension $d = 4$ in the Lagrangean case or $d = 3$ in the Eulerian case; S^d contains C^1. In contrast, the parabolic orbits form the center-unstable manifold U^d (no exponents > 0, Lagrange: $d = 5$, Euler: $d = 6$). U^d contains C^1 as well; its orbits behave like outgoing orbits emerging from points of C^1. The phase space trajectory of a close triple encounter first follows S^d, then transfers to the neighbourhood of U^d.

3.2   The triple-collision manifold

In this final section we will give an outlook on the concept of the triple-collision manifold, introduced by McGehee in 1974 [10]. Although McGehee's original paper only discusses the 1-dimensional three-body problem it is the key for rigorously handling many more complicated cases of multiple-collision singularities. In the sequel the isosceles triple encounter was discussed by Simó [17], Devaney [3], and Irigoyen [9]; the general planar triple encounter was tackled by Hulkower [8], Moeckel [12], and this author [28], [29] = [30], to mention just a few. The concept of the collision manifold had been anticipated in 1971 by Easton [6]; it was generalized later by Devaney [4], McGehee [11].

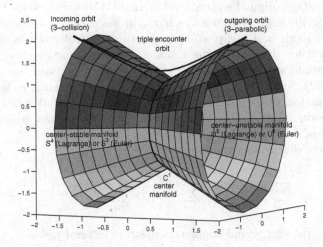

Fig. 5. Qualitative sketch of the invariant manifolds in phase space $\mathbb{R}^8$

The basic idea is to apply the scaling transformation (46) in the form

$$x_j = r\, z_j, \;\; p_j = r^{-\frac{1}{2}}\, u_j; \;\; dt = r^{\frac{3}{2}}\, ds, \;\; j = 1, 2, 3 \tag{50}$$

to the equations of motion (3), where the scaling factor r is the **variable** quantity satisfying

$$r^2 = I = \sum_{k=1}^{3} m_k\, |x_k|^2, \tag{51}$$

i.e. r is the radius of inertia of the three-body system. From (50_3) there follows the differentiation rule $d/dt = r^{-3/2} \cdot d/ds$. By denoting differentiations with respect to the new independent variable s by primes we obtain from (51):

$$r' = r \cdot \sum_{k=1}^{3} z_k\, u_k. \tag{52}$$

Furthermore, the equations of motion (3) are transformed into

$$
\begin{aligned}
z_j' &= -z_j \sum_{k=1}^{3} z_k\, u_k + \frac{u_j}{m_j}, \\
u_j' &= \frac{u_j}{2} \sum_{k=1}^{3} z_k\, u_k + \frac{\partial}{\partial z_j}\, U(z), \;\; j = 1, 2, 3.
\end{aligned}
\tag{53}
$$

The system (52), (53), together with

$$t' = r^{\frac{3}{2}}, \tag{54}$$

is equivalent to the original equations of motion (3). However, since the new co-ordinates describe triple collisions in a blown-up fashion, transition into collision solutions is smooth, and it is possible to describe near-collisions as well.

The reason for this is that the collision solutions, characterized by $r = 0$, form an invariant submanifold of the manifold of all solutions, i.e. $r = 0$ is a solution of (52). For solutions with $r = 0$ time does not advance, as follows from (54). The manifold of the solutions of the reduced system (53) is defined as the **triple-collision manifold** $\mathcal{T}$. The flow on $\mathcal{T}$ bears no physical reality; it is the limiting flow of an arbitrarily sharp triple close encounter.

The system (53) admits the integrals

$$\sum_{j=1}^{3} m_j z_j = 0, \quad \sum_{j=1}^{3} u_j = 0, \quad \sum_{j=1}^{3} m_j z_j^2 = 1. \tag{55}$$

Transforming the energy and angular momentum integral yields

$$rH = \frac{1}{2} \sum_{j=1}^{3} \frac{u_j^2}{m_j} - U(z), \quad C\, r^{-\frac{1}{2}} = \sum_{j=1}^{3} z_j * y_j. \tag{56}$$

Therefore, on $\mathcal{T}$ we necessarily have $rH = 0$ and $C = 0$, but in general $\sum z_j * u_j \neq 0$. $\mathcal{T}$ is imbedded in 12-dimensional phase space. The integrals (55) and the energy integral $rH = 0$ in (56) reduce $\mathcal{T}$ to a 6-dimensional manifold. Since the equations (53) are invariant under orthogonal rotations $z_j \longmapsto Rz_j$, $u_j \longmapsto Ru_j$ ($R^T R = Id$) a further reduction of the dimensionality may be achieved.

The equilibrium points E_k of the flow on $\mathcal{T}$ play an important role since E_k are the points where the orbits in phase space can approach or leave $\mathcal{T}$. It may be shown that the points E_k correspond to the central configurations of the three-body system, where each of them gives rise to two equilibria, one for collision solutions, the other for explosion solutions.

The phase space orbit of a close encounter solution will therefore approach $\mathcal{T}$ near an equilibrium E_k. Then the orbit remains near $\mathcal{T}$ either until it evolves into a hyperbolic-elliptic escape. This means the projection of the orbit onto $\mathcal{T}$ spirals around pointlike "holes" of $\mathcal{T}$. Alternatively (rarely!), the orbit near $\mathcal{T}$ may approach another equilibrium, E_ℓ, and leave the vicinity of $\mathcal{T}$ in an inverse triple collision, a triple explosion.

References

1. J.D. Cole: *Perturbation Methods in Applied Mathematics.* (Ginn-Blaisdell 1968). Also: J. Kevorkian, J.D. Cole : *Perturbation Methods in Applied Mathematics.* Applied Mathematical Sciences **34**, Springer (1981), 558 pp.
2. J. Chazy: *Sur l'allure du mouvement dans le problème des trois corps quand le temps croît indéfiniment.* Ann. Sci. Ecole Norm. **39** (1922), pp. 29-130.
3. R.L. Devaney: *Triple collision in the planar isosceles three body problem.* Inventiones mathematicae **60** (1980), pp. 249-267.

4. R.L. Devaney: *Blowing up singularities in classical mechanical systems.* American Mathematical Monthly **89** (8) (1982), pp. 535-552.
5. R. Easton: *Regularization of vector fields by surgery.* J. of Diff. Equations, **10**(1) (1971), pp. 92-99.
6. R. Easton: *Some topology of the 3-body problem.* J. of Diff. Equations, **10**(2) (1971), pp. 371-377.
7. D. Gruntz, J. Waldvogel: 'Orbíts in the planar problem of three bodies'. In: *Solving Problems in Scientific Computing Using Maple and MATLAB*, eds. W. Gander and J. Hrebicek, Springer (1993), pp. 37-57.
8. N.D. Hulkower: *The zero energy three body problem.* Indiana University Math. Journal **27**(3) (1978), pp. 409-447.
9. J.M. Irigoyen: *La variété de collision triple dans le cas isoscèle du problème des trois corps.* C.R. Acad. Sci. Paris **290** (1980).
10. R. McGehee: *Triple collision in the collinear three-body problem.* Inventiones math. **27** (1974), pp. 191-227.
11. R. McGehee: *Singularities in classical celestial mechanics.* Proceedings of the International Congress of Mathematicians, Helsinki (1978), pp. 827-834.
12. R. Moeckel: *Orbits near triple collision in the three-body problem.* Forschungsinstitut für Mathematik, ETH Zürich (1981).
13. C.L. Siegel: *Der Dreierstoß.* Ann. Math. **42** (1941), pp. 127-168.
14. C.L. Siegel: *Lectures on the singularities of the three-body problem.* Tata Institute, Bombay (1967).
15. C.L. Siegel, J.K. Moser: *Lectures on Celestial Mechanics.* Springer (1971), 290 pp.
16. C. Simó: *Masses for which triple collision is regularizable.* Presented at the meeting "Mathematical Methods in Celestial Mechanics", Oberwolfach (1978).
17. C. Simó: 'Analysis of triple collision in the isosceles problem'. In: *Classical Mechanics and Dynamical Systems*, Marcel Dekker (1981).
18. K. Sundman: Mémoire sur le problème des trois corps. Acta Math. **36** (1912), pp. 105-179.
19. J. Waldvogel: *Note concerning a conjecture by A. Wintner,* Celestial Mechanics **5** (1972), pp. 37-40.
20. J. Waldvogel: *A new regularization of the planar problem of three bodies.* Celestial Mechanics **6** (1972), pp. 221-231.
21. J. Waldvogel: 'Collision singularities in gravitational problems'. in: *Recent Advances in Dynamical Astronomy*, eds. V. Szebehely and B.D. Tapley, Reidel (1973), pp. 21-33.
22. J. Waldvogel: *Altes und Neues über das Dreikörperproblem,* Verhandlungen der Schweizerischen Naturforschenden Gesellschaft (1974), pp. 77-81.
23. J. Waldvogel: *The close triple approach.* Celestial Mechanics **11** (1975), pp. 429-432.
24. J. Waldvogel: 'Triple collision'. In: *Long Time Prediction in Dynamics*, eds. V. Szebehely and B.D. Tapley, Reidel (1976), pp. 241-258
25. J. Waldvogel: *The three-body problem near triple collision.* Celestial Mechanics **14** (1976), pp. 287-300.
26. J. Waldvogel: *Triple collision as an unstable equilibrium.* Bull. Acad. Royale de Belgique, Classe des Sci. **63** (1977), pp. 34-50.
27. J. Waldvogel: 'Stable and unstable manifolds in planar triple collision'. In: *Instabilities in Dynamical Systems*, ed. V. Szebehely, Reidel (1979), pp. 263-271.
28. J. Waldvogel: *La variété de collision triple.* C.R. Acad. Sc. Paris **288** (1979), pp. 635-637.

29. J. Waldvogel: *Symmetric and regular coordinates on the planar triple collision manifold.* Celestial Mechanics **28** (1982), pp. 69-82.

30. J. Waldvogel: 'Coordonnées symétriques sur la variété de collision triple du problème plan des trois corps'. In: *Applications of Modern Dynamics to Celestial Mechanics and Astrodynamics*, ed. V. Szebehely, Reidel (1982), pp. 249-266.

31. A. Wintner: *The Analytical Foundations of Celestial Mechanics.* Princeton University Press (1941), 448 pp.

Dynamical and Kinetic Aspects of Collisions

Yves Elskens

Equipe turbulence plasma, Laboratoire de physique des interactions ioniques et moléculaires (UMR 6633 CNRS–Université de Provence),
case 321, campus Saint-Jérôme, F-13397 Marseille cedex 20
elskens@newsup.univ-mrs.fr

Abstract. For deterministic dynamics of a finite number of interacting bodies, collisions are brief events sensitive to initial conditions. They may contribute to the chaotic evolution of the system at microscopic scale.

The limit of $N \to \infty$ bodies requires a different formulation, kinetic theory, in which the dynamics leads to the BBGKY hierarchy. Two limits enable one to reduce the hierarchy to a single partial differential equation for a one-particle distribution function, respectively the Vlasov equation and the Boltzmann equation. In the mean-field limit, individual binary interactions are negligible. To the contrary, in the Boltzmann–Grad limit, individual collisions are crucial and invite to a stochastic description of particle motion.

Recent results provide a better insight on the relaxation to equilibrium for the Boltzmann equation.

1 Introduction

There are as many ways to solve a problem in mechanics as there are types of questions in which one is actually interested. In particular the discussion of collisions takes different forms depending on whether one focuses on few bodies or many bodies.

As the Newton potential is specially difficult to discuss, results will be quoted which apply to its repulsive counterpart (Coulomb potential) and to even simpler interactions such as the hard sphere elastic collisions, which emphasize the short-time aspect in collisions. We focus on the hamiltonian models of interacting identical bodies. The case of a small body in the field of large ones has also been much studied, and its prototype is the Lorentz gas with hard elastic collisions [3,5].

The results summarized in this contribution are discussed more thoroughly in the textbooks by Balescu [2], Dorfman [3] and Spohn [5]. More recent advances in the theory of kinetic equations are drawn from Villani's works [6,7]. Though an accurate (i.e. correct) formulation of the results must be quite mathematical, the author would prefer the reader to focus on the essence of the statements rather than on technical issues.

2 N-body dynamics

The dynamics for N bodies (or particles in this contribution) is formulated in phase space $\Gamma = \gamma^N$. Each body (labeled $1 \leq j \leq N$) is described as a 'structureless point', with position $q_j \in \Lambda \subset I\!\!R^d$ (with $d = 3$ usually) and momentum $p_j \in I\!\!R^d$, so that each particle is represented by a point in Boltzmann's μ-space $\gamma = \Lambda \times I\!\!R^d$ and the full state of the system at any time t is $x = (q_1, p_1, \ldots q_N, p_N)$. The equations of motion are first-order ordinary differential equations

$$\dot{x} = w(x) \tag{1}$$

where the vector field $w : \Gamma \to I\!\!R^{2Nd}$ describes both velocities $\dot{q}_j$ and forces $\dot{p}_j$. In the familiar hamiltonian case of N identical bodies of mass m interacting through a pair potential U, the vector field w derives from a hamiltonian

$$H : \Gamma \to I\!\!R : x \mapsto H(x) = \sum_j \frac{|p_j|^2}{2m} + \frac{1}{2} \sum_{i \neq j} U(q_i - q_j) \tag{2}$$

generating the equations of motion

$$\dot{q}_j = p_j/m \tag{3a}$$
$$\dot{p}_j = -\sum_{i \neq j} \nabla_q U(q_j - q_i) \tag{3b}$$

The first question raised by these equations is the **existence of the dynamics**, i.e. whether, given admissible initial data $x \in \Gamma$ at time 0, the system (1) or (3a-3b) admits a solution in Γ for all times $t \geq 0$. This is the Cauchy problem, and the possibility of collisions ($q_i = q_j$ for some $1 \leq i \neq j \leq N$) may be a difficulty. The trouble is obvious if bodies do actually collide, but also for the accurate computation of their motion if near-collisions occur, as discussed in other contributions in this volume.

Repelling forces can be controlled by requiring total energy to be finite. The finiteness of the force itself can be controlled by considering a short-range interaction (decaying fast enough for $|q| \to \infty$) and ensuring that there are not too many particles in a finite-size domain. The uniqueness of the solutions is ensured by a smooth enough potential. Technically, one proves e.g. (see Spohn [5], p. 12) that, for $N \to \infty$ in an infinite domain Λ, the Cauchy problem admits a solution for all times for almost any initial data if U is either (i) the hard sphere potential or (ii) a three times continuously differentiable, finite range, superstable potential.

Though gravitation fails to meet these requirements, the study of potentials fulfilling them is interesting and may help in discussing the gravitational case by means of approximations.

3   Invariants, approximate motion and collisions

Integrating the equations of motion (3a-3b) means finding the flow $\Theta_t : \gamma^N \to \gamma^N : x(0) \mapsto x(t)$ for all times t. This is generally too demanding, and one develops insight by finding special solutions (such as stationary states or periodic solutions) and reducing the domain in which a solution may wander.

In this respect, first integrals are crucial. Their conservation reduces the dimension of the manifold accessible to each solution, and they provide tests for the accuracy of the numerical integration of the trajectories. In the extreme case of integrable dynamics, their use reduces the integration of the motion to a mere sequence of changes of variable from (q_j, p_j) to action-angle variables (I_i, φ_i), along with integrating the free evolution of the angles in the form $\varphi_i(t) = \varphi_i(0) + \Omega_i(I)t$.

Integrable systems are exceptions, and one usually resorts to approximations. Almost-invariants may often be identified, taking advantage from the fact that typically the right-hand side of (3b) for a given particle j may be small over some long times and be significantly large for short 'bursts'. Such strong-force events are the collisions. The central idea is the **separation of two scales**:

1. the 'long' time scale, over which a body moves essentially freely over a large distance, with almost constant momentum, and
2. the 'short' time scale, over which a body suffers a significant impulse, with little change in its position.

Whereas the duration of a collision is controlled by the interaction dynamics, the intercollision time is controlled by the spatial density of the bodies, their typical velocity and the range of the interaction. In this respect, collisions and free motion are best disentangled in a 'dilute' system.

For short-range forces, the collisions may involve only two bodies at a time, if the bodies do not bind in binary systems. This is a major simplification, and one may then describe each collision quite accurately, as a mapping from incoming (q_1, p_1, q_2, p_2) to outgoing (q_1', p_1', q_2', p_2'). This applies rigorously to elastic hard spheres, but also reasonably to smoother interactions, including attractive ones (if no binaries are formed).

It must also be noted that generally dynamics in many-body systems is chaotic [2,3]. Therefore, small errors on initial data will grow in time, and the characteristic rate for their exponential growth defines the Lyapunov exponents. Considering two initial conditions, close to each other, say $x(0)$ and $x(0) + \delta x(0)$, the dynamics evolves two trajectories $x(t)$ and $x(t) + \delta x(t)$, and the largest Lyapunov exponent associated to the trajectory $x(t)$ is

$$\lambda(x(0)) = \lim_{t \to \infty} \limsup_{|\delta x(0)| \to 0} \frac{1}{t} \ln \frac{|\delta x(t)|}{|\delta x(0)|} \tag{4}$$

where the limit with respect to $\delta x(0)$ accounts for the fact that the infinitesimal initial perturbation may be in any direction in phase space. It is easily shown that $\lambda(x(0)) = \lambda(x(t_0))$ for any t_0, i.e. the largest exponent does not depend on the specific initial time from which one integrates the motion, but it generally depends on the trajectory itself.

Chaos on the infinitesimal scale manifests itself by $\lambda(x(0)) > 0$. It may happen that λ be independent of the trajectory too ; this is discussed in the context of ergodic theory.

The chaoticity of dynamics invites one to investigate not just the fate of the trajectory emanating from a single initial condition, but also the fate of nearby trajectories. This is one motivation for a statistical physics approach to many-body dynamics.

4   Collisions and Lyapunov exponents

Two-body collisions between bodies with spherical symmetry (interacting by a central potential) conserve the two bodies (hence total mass), total momentum, total energy and total angular momentum. Assuming that the interaction potential and force vanish (or are negligible) when bodies are far enough, the outgoing momenta are obtained from the incoming ones using only the relative velocity $u = (p_2 - p_1)/m$ and the impact parameter $b = |J|/|mu|$ where J is the orbital momentum (in the center-of-mass reference frame).

For a system with collisions, the free-motion intervals as well as the collision events contribute to the separation of trajectories. During the free motion, the trajectories do not diverge exponentially from each other but only linearly in time:

$$\begin{pmatrix} \delta q_j(t) \\ \delta p_j(t) \end{pmatrix} = \begin{pmatrix} 1 & t \\ 0 & 1 \end{pmatrix} \cdot \begin{pmatrix} \delta q_j(0) \\ \delta p_j(0) \end{pmatrix} \tag{5}$$

On the other hand, 'impulsive' collisions do not change the positions significantly and thus do not enhance the separation of trajectories in the position space, but they strongly affect the momentum components. It is the succession of these events, coupled with the linear increase of separation in free motion, which ultimately causes the average exponential separation of trajectories measured by the positive Lyapunov exponent.

The momentum part of the collision equations reads[1]

$$p_1' = \frac{1}{2}(p_1 + p_2 + mu') = p_1 + \frac{m}{2}(u' - u) \tag{6a}$$

$$p_2' = \frac{1}{2}(p_1 + p_2 - mu') = p_2 - \frac{m}{2}(u' - u) \tag{6b}$$

where the post-collision asymptotic relative velocity u' is in the plane of b and u, and $|u'| = |u|$. The collision is just a deflection of the relative velocity, with an angle θ which depends on b, on $|u|$ and on the specific interaction potential. For technical purposes, it may be convenient to parametrize the collision by $(|u|, \omega)$ rather than by (u, b), where ω is the unit vector in the direction of closest approach (given by $u' - u$). As the collision map is not linear, its jacobian matrix $\mathcal{C}$ is not constant and, in general, it may have an eigenvalue with modulus larger than 1.

---

[1] This system must be completed by equations for the positions, notably for the outgoing impact parameter b', which also satisfies $|b'| = |b|$ and b' coplanar with b and u.

The long time evolution of the small perturbations to a trajectory is thus represented by expressions like

$$\delta x(t) = \frac{\partial x(t)}{\partial x(0)} \cdot \delta x(0) = \frac{\partial \Theta_t(x(0))}{\partial x(0)} \cdot \delta x(0) \tag{7}$$

$$= \begin{pmatrix} 1 & t - t' \\ 0 & 1 \end{pmatrix} \cdot \mathcal{C}(t') \cdot \begin{pmatrix} 1 & t' - t'' \\ 0 & 1 \end{pmatrix} \ldots \delta x(0) \tag{8}$$

where the $2Nd \times 2Nd$-dimensional jacobian matrices $\mathcal{C}$ describe the successive collisions. In general, these matrices do not commute with each other, nor do they commute with the free motion matrices. The resulting product $\partial \Theta_t(x(0))/\partial x(0)$ is likely to have some eigenvalues with modulus larger than one, and one may try to estimate the largest one in the limit $N \to \infty$, $t \to \infty$ using a (noncommutative) central limit theorem for random matrices. But such a central limit theorem requires that successive collisions be (almost) independent, which is a very strong assumption: once two bodies have interacted, knowledge of the past motion of one of them provides information on the motion of the other one ; hence the two motions are not independent. Therefore one must argue that successive collisions will involve new partners, and the sensitivity of the collision outcome to small perturbations may help to decrease the correlation in time between the bodies [2,3].

In the case of hard spheres, one can discuss the separation of nearby trajectories using geometrical optics (see Dorfman [3], ch. 18): the instantaneous collision reflects the colliding body in the same way as a spherical mirror reflecting a light ray, and the free motion between collisions is equivalent to the free propagation of light. Nearby trajectories may then be considered as a light front, and the convexity of the spherical mirrors causes the dispersion of the light front. One can then estimate the largest Lyapunov exponent of the system, and find it positive indeed.

5   Kinetic theory and BBGKY hierarchy

Kinetic theory is the relevant way to discuss the many-body dynamics in terms of few-body interactions and motion. The first unknown in the kinetic model is the actual number N of bodies to evolve in the space $\gamma = \Lambda \times \mathbb{R}^d$. Then, given N, the initial microstate x may be considered as a random variable in γ^N with a probability density[2] f_N (up to normalisation).

If one is interested in the joint distribution of any n particles in γ, the relevant probability density on γ^n is the correlation function

$$\rho_n(q_1, p_1, \ldots, q_n, p_n) = \\ \sum_{k=0}^{\infty} \frac{1}{k!} \int_{\gamma^k} f_{n+k}(q_1, p_1, \ldots, q_{n+k}, p_{n+k}) \prod_{s=1}^{k} dq_{n+s} dp_{n+s} \tag{9}$$

The n-body densities f_n and the correlation functions determine each other uniquely, as $f_n = \sum_{k=0}^{\infty} \frac{(-1)^k}{k!} \int_{\gamma^k} \rho_{n+k}$.

[2] The f_N are symmetric, positive and normalized so that $\rho_0 = 1$ in (9).

The case of independently distributed particles occurs when correlation functions factorize as $\rho_n(q_1, p_1, \ldots, q_n, p_n) = \prod_{k=1}^{n} \rho_1(q_k, p_k)$. In that case, the number $\nu(\Delta)$ of particles with position and momentum (q, p) in a domain $\Delta \subset \gamma$ follows a Poisson distribution with expectation $N' = \int_\Delta \rho_1(q, p) \mathrm{d}q \mathrm{d}p$, i.e. $\mathbb{P}(\nu(\Delta) = n) = (N'^n/n!)\mathrm{e}^{-N'}$.

A n-body observable is a function $g_n : \gamma^n \to \mathbb{R}$, continuous with compact support (or a limit of such functions). For a statistical state with correlation functions ρ_n, one computes the expectation of g_n directly as

$$\langle g_n \rangle = \frac{1}{n!} \int_{\gamma^n} g_n \rho_n \prod_{s=1}^{n} \mathrm{d}q_s \mathrm{d}p_s \tag{10}$$

The evolution equations (3a-3b) are easily translated into evolution equations for the n-particle distribution functions and correlation functions. The results are the n-body Liouville equations for the f_n and the Bogoliubov-Born-Green-Kirkwood-Yvon (BBGKY) hierarchy for the ρ_n. The latter reads formally

$$\partial_t \rho_n - L_n \rho_n = C_{n,n+1} \rho_{n+1} \tag{11}$$

with the collision integral taking the form

$$C_{1,2}\rho_2 = \int_\gamma \delta(|Q| - a)\, Q \cdot \frac{P}{m}\, \rho_2(q_1, p_1, q_1 + Q, p_1 + P)\mathrm{d}Q\mathrm{d}P \tag{12}$$

in the case of elastic hard spheres with diameter a. In this hierarchy,

$$L_n \rho_n = -\frac{p}{m} \cdot \nabla_q \rho_n \tag{13}$$

is the free motion (streaming) operator, describing noninteracting particles. The system of equations (11) is a hierarchy, as it expresses the evolution of the n-th function ρ_n in the sequence of correlation function in terms of ρ_{n+1}, the next function of this sequence ; hence, the determination of one ρ_k is intimately related with the determination of all ρ_n, $n \geq 1$.

The free motion is integrable. A vanishing right-hand side in (11) yields

$$\rho_n(q_1, p_1, \ldots q_n, p_n, t) = \mathrm{e}^{L_n t} \rho_n(q_1, p_1, \ldots q_n, p_n, 0)$$
$$= \rho_n(q_1 - (p_1/m)t, p_1, \ldots q_n - (p_n/m)t, p_n, 0) \tag{14}$$

Introducing the free streaming operator $S_n(t) = \mathrm{e}^{L_n t}$ into the original system (11), the BBGKY hierarchy is formally solved by the collision expansion

$$\rho_n(t) = S_n(t)\rho_n(0) + \int_0^t S_n(t - t')C_{n,n+1}\rho_{n+1}(t')\mathrm{d}t' \tag{15}$$

$$= S_n(t)\rho_n(0) + \int_0^t S_n(t - t')C_{n,n+1}\rho_{n+1}(0)\mathrm{d}t'$$

$$+ \int_0^t \int_0^{t'} S_n(t - t')C_{n,n+1}S_{n+1}(t' - t'')C_{n+1,n+2}\rho_{n+2}(0)\mathrm{d}t''\mathrm{d}t'$$

$$+ \ldots \tag{16}$$

However, extracting the long-time evolution of correlation functions from this expansion is awkward. Two simplifying approximations have been fruitfully investigated, the mean-field limit for long-range interactions and the Boltzmann–Grad limit for short-range interactions.

6   Mean-field limit and Vlasov equation

The mean field limit, or the weak coupling limit, is appropriate if the interaction is long range. Such interactions are best modeled by hamiltonians in the form

$$H_N(q,p) = \sum_j \frac{|p_j|^2}{2m} + \frac{1}{2N} \sum_{i \neq j} U(q_i - q_j) \tag{17}$$

where the coupling constant is scaled by N to ensure that the energy is extensive in the limit $N \to \infty$ and that the contribution of each individual particle to the total force is small in this limit. In the limit $N \to \infty$, the one-particle correlation function describing the microstate of the system may converge to a (hopefully smooth) distribution on γ,

$$\rho_1(q,p) = \frac{1}{N} \sum_{j=1}^{N} \delta(q_j - q)\delta(p_j - p) \underset{N \to \infty}{\longrightarrow} \rho_1^*(q,p) \tag{18}$$

The space of finite measures on γ, denoted by $L^1(\gamma)$, is endowed with a norm $|.|$ called the bounded Lipschitz norm[3]. One expects the distribution to obey a kinetic equation, and this is well proved if the interaction is not too singular (see e.g. Spohn [5], ch. 5):

Theorem: *If $\nabla_q U$ is bounded and Lipschitz continuous, then*

1. *For any initial measure $\rho_1(x,0)$ in $L^1(\gamma)$, the Vlasov equation*

$$\partial_t \rho_1(q,p,t) + \frac{p}{m} \cdot \nabla_q \rho_1(q,p,t) + E_1(q,t) \cdot \nabla_p \rho_1(q,p,t) = 0 \tag{19}$$

   coupled with the self-consistent mean field

$$E_1(q,t) = -\int_\gamma [\nabla_q U(q-q')]\rho_1(q',p',t)\mathrm{d}q'\mathrm{d}p' \tag{20}$$

   has a unique solution in $L^1(\gamma)$ for all times $t \geq 0$.
2. *Consider two solutions $\rho_1(q,p,t)$, $\sigma_1(q,p,t)$ of the Vlasov equation (each with its associated field evolution). Then*

$$|\rho_1(.,.,t) - \sigma_1(.,.,t)| \leq e^{ct}|\rho_1(.,.,0) - \sigma_1(.,.,0)| \tag{21}$$

   for some constant c (independent of ρ_1 and σ_1).

---

[3] This norm is defined in terms of the observables on γ [5].

The constant c depends only on the form of the potential U and is related to two constants B and L which occur naturally. Here, B is an upper bound on the actual force acting on any particle at any time ($|\nabla_q U| \leq B$), and L is an upper bound on any force gradient (Lipschitz continuity: $|\nabla_q U(q) - \nabla_q U(q')| \leq L|q - q'|$). A hint to the theorem is the fact that the Lyapunov exponents for the trajectories of the N-body problem may be estimated from the relation

$$\begin{pmatrix} \delta\dot{q} \\ \delta\dot{p} \end{pmatrix} = \begin{pmatrix} 0 & 1/m \\ \sim L & 0 \end{pmatrix} \cdot \begin{pmatrix} \delta q \\ \delta p \end{pmatrix} \tag{22}$$

independently of N. Then one also notes that increasing the number of particles in the system for the mean-field limit does not change the force on each particle very much, because each of them contributes only with a factor $1/N$ to the total force (one may interpret (18) as an application of the Riemann integral). Thus, considering more particles does not dramatically affect the dynamics in the system, and the existence of the finite-N dynamics leads to the solution of the limiting kinetic equation.

One further shows that, under natural hypotheses for the initial data, the correlation functions are factorized in the weak coupling limit as $N \to \infty$. This validates rigorously the use of test particles in the field $E_1(q,t)$ to follow the actual motion of typical particles. It is also clear that the kinetic equation provides a correct description of the many-body system, and integrating the partial differential equation (19) coupled with the integral field equation (20) provides a reasonable alternative to the full dynamics of the N-body system.

However, the mean-field kinetic limit theorem above has two drawbacks. First, the exponential estimate (21) allows for a rather fast deterioration of the accuracy η on the distribution ρ_1. If $\eta_t < e^{ct}\eta_0$, then $\eta_0 > e^{-ct}\eta_t$; to be on the safe side and to ensure η_t at time t, one will require an initial accuracy $e^{-ct}\eta_t$, which goes to 0 exponentially for increasing t.

In other words, given an initial accuracy η_0, an accuracy η_t can be ensured only for times $t \leq c^{-1} \ln \frac{\eta_t}{\eta_0}$. As good representations of distributions by a finite number of points yield $\eta_0 \sim N^{-1}$, the computations of finite-N evolution and the Vlasov kinetic evolution may agree (within given η_t) only over rather short (increasing only as $\ln N$) time scales.

The second drawback relates to the collisionless nature of the Vlasov equation. In γ space, it is naturally interpreted as a transport equation for the distribution ρ_1, i.e. the latter is convected by the vector field $(p/m, E_1(q,t))$. It turns out that the latter is conservative, i.e.

$$\text{div}_{(q,p)}(p/m, E_1) = \nabla_q \cdot (p/m) + \nabla_p \cdot E_1(q,t) = 0 \tag{23}$$

Consider now the γ-space domain below a given level α for the density ρ_1, i.e. $A(\alpha,t) = \{(q,p) : \rho_1(q,p,t) < \alpha\}$: one shows that $\frac{d}{dt}A(\alpha,t) = 0$. Similarly, the functional $\mathcal{H}(\rho,t) = \int_\gamma \rho(q,p,t) \ln \rho(q,p,t) dq dp$ and all $\mathcal{A}_s(\rho,t) = \int_\gamma [\rho(q,p,t)]^s dq dp$ are conserved for all time t. This is surprising in view of the usual relaxation of the distribution of many bodies to some equilibrium.

The mathematical solution to this paradox is that the approach to equilibrium is described in coarse ways: e.g. one monitors the time evolution of just one (or a few) smooth observable(s) g. Then it suffices that the actual distribution ρ, which is not constant, exhibit oscillations on increasingly smaller scales as t increases: as the smooth g averages off these oscillations, the evolution of ρ will appear as leading towards equilibrium. This problem of mixing is discussed in the books by Balescu [2] and Dorfman [3].

7 Vlasov–Poisson equation for Coulomb and Newton interactions

The Coulomb potential does not satisfy the hypotheses of the previous theorem. However, a similar result holds [5], so that the Vlasov–Poisson system is a genuine mean-field limit of the N-body dynamics for this interaction too ; Newton potential and vortex dynamics in a 2-dimensional fluid are also discussed [5]. The long-time evolution of the Vlasov–Poisson system is presently a matter of debate.

This is somewhat related to the question of the self-force in the dynamics of point particles: the point particle equation of motion $\dot{p}_j = F(q_j, t)$ makes sense only if $\lim_{q \to q_j} F(q)$ exists. However, as the Newton force field generated by particle j diverges like $|q - q_j|^{-2}$ as $q \to q_j$, one must define a 'regular' limit for the field. A convenient way to bypass the difficulties of the newtonian short-range singularity is to replace the actual potential by a smoother one, like $U_\varepsilon(q) = -Gm(|q|^2 + \varepsilon^2)^{-1/2}$ and let $\varepsilon \to 0$ only after all calculations are finished. More elaborate approaches may use self-consistent soliton-like solutions to the kinetic equations, as was done by Appel and Kiessling for the Maxwell equations [1].

Of interest to astronomy is also the discussion of the relativistic version of the Vlasov–Poisson system. For this relativistic version, one takes into account the velocity-momentum relation $\dot{q} = p/\sqrt{m^2 + |p|^2 c^{-2}}$, which is substituted for p/m in (19). Then, if one neglects the retardation effects in the field propagation (i.e. one keeps (20), rather than replace it by a wave equation), the total relativistic energy reads

$$H = \int_\gamma \left(\sqrt{m^2 c^4 + p^2 c^2} + \frac{m\mathcal{U}(q)}{2} \right) \rho_1(q, p) \mathrm{d}p \mathrm{d}q$$
$$= \int_\gamma \sqrt{m^2 c^4 + p^2 c^2} \rho_1(q, p) \mathrm{d}p \mathrm{d}q - \frac{1}{8\pi G} \int_{I\!R^3} |\nabla \mathcal{U}(q)|^2 \mathrm{d}q \qquad (24)$$

with the self-consistent potential $U = m\mathcal{U}$ satisfying the Poisson equation,

$$\nabla^2 \mathcal{U}(q) = 4\pi Gm \int_{I\!R^3} \rho_1(q, p) \mathrm{d}p \qquad (25)$$

where G is the gravitation constant. Glassey and Schaeffer [4] have shown that, if $H < 0$ (i.e. if the 'non-relativistic energy' $H - mc^2 \int_\gamma \rho_1(q, p) \mathrm{d}p \mathrm{d}q$ is 'very negative'), then the solutions to (19)-(20) with spherically symmetric initial conditions do not exist for all times, i.e. the system evolves to a **singularity in a finite time**.

Finite-N corrections to the mean-field limit are a difficult chapter of kinetic theory [2,5]. The dominant correction to the Vlasov–Poisson system is formulated in terms of the Balescu-Lenard equation. While the Vlasov equation conserves entropy, the Balescu-Lenard equation admits a H-theorem like the Boltzmann equation.

8   Boltzmann–Grad limit and Boltzmann equation

For short-range forces and in the low-density limit, the Boltzmann equation has been proved to yield the correct evolution of the correlation functions over a finite (unfortunately short) time interval. Its fundamental assumption, the **molecular chaos** hypothesis, is to use the Ansatz $\rho_2(q_1, p_1, q_2, p_2) = \rho_1(q_1, p_1)\rho_1(q_2, p_2)$ in the BBGKY hierarchy. One may indeed take a set of factorized correlation functions as initial conditions for the hierarchy, but in general the factorization does not remain valid after the initial time: Boltzmann's assumption of **propagation** of molecular chaos does not hold in general. However, the free streaming operator does propagate molecular chaos: only the collisions can (and do indeed) break down the independence between the bodies.

We denote the Maxwell-Boltzmann velocity distribution function by

$$M_T(v) = (2\pi T/m)^{-d/2} e^{-m|v|^2/(2T)} \tag{26}$$

where we set the Boltzmann constant $k_B = 1$.

To control the recollisions (i.e. make their effect negligible in some limit), Lanford took advantage of the Grad scaling, in which the collision cross section goes to zero and the density of bodies goes to infinity, while the mean free path remains fixed. The rescaled correlation functions are $r_n^\varepsilon = \varepsilon^{2n}\rho_n$ for hard spheres with rescaled diameter $a^\varepsilon = \varepsilon a$. Then (see e.g. Spohn [5], ch. 4)

Theorem: *Consider a sequence of initial correlation functions r_n^ε for finite ε such that*

1. *there exist $C, T, z > 0$ such that, for any $n > 0$, $|r_n^\varepsilon(q_1, p_1, \ldots q_n, p_n, 0)| \leq C \prod_{j=1}^n \left(z M_T(p_j/m)\right)$, and*
2. *there exist correlation functions $r_n : \gamma^n \to \mathbb{R}$ such that $\lim_{\varepsilon \to 0} r_n^\varepsilon(.,.,0) = r_n(.,.,0)$.*

Then

1. *there exists a time $t_0 > 0$ such that, for all $0 \leq t < t_0$, the rescaled correlations functions solutions of the BBGKY hierarchy $r_n^\varepsilon(.,.,t)$ admit limits $r_n(.,.,t) = \lim_{\varepsilon \to 0} r_n^\varepsilon(.,.,t)$.*

2. *Moreover, if initially $r_n(q_1, p_1, \ldots q_n, p_n, 0) = \prod_{j=1}^{n} r_1(q_j, p_j, 0)$, then the factorization propagates: $r_n(q_1, p_1, \ldots q_n, p_n, t) = \prod_j r_1(q_j, p_j, t)$, and the one-particle function satisfies the Boltzmann equation*

$$\partial_t r_1 + \frac{p}{m} \cdot \nabla_q r_1 = C_{1,2}(r_1 r_1) \tag{27}$$

This theorem is encouraging, as it formally provides a rigorous foundation to the use of the Boltzmann equation in kinetic theory. However, the critical time t_0 is only a fraction of the mean intercollision time τ (about $\tau/5$ for hard spheres in $d = 3$ dimensions, and $t_0 \to 0$ as $d \to \infty$). Larger estimates for t_0 have been achieved for systems which expand in vacuum: particles move away from each other, and would naturally have fewer collisions in the future. The difficulty to overcome, in order to increase t_0, is not only due to the complexity of the recollisions to be analysed but also to the mathematical difficulties inherent to the Boltzmann equation.

9   Entropy dissipation for the Boltzmann equation

Recent progress in the theory of the Boltzmann equation provides estimates on the rate of approach to equilibrium of its solutions. The results do not extend to Newton potential (!), but are nevertheless quite instructive.

Let us write the Boltzmann equation for the one-particle density $f(q, v, t) = m\rho_1(q, mv, t)$ in the form

$$(\partial_t + v \cdot \nabla_q)f = \int_{\mathbb{R}^d} \int_{S^{d-1}} B(v - v_*, \omega)(f'f'_* - ff_*) \, d\omega dv_* \tag{28}$$

where v_* is the velocity of the collision partner and ω is the direction of the vector of shortest approach of the particle by its partner[4].

Consider for simplicity a collision differential cross section of the form

$$B(u, \omega) = |u|^\kappa \beta(\cos \theta) \tag{29}$$

where $u \cdot \omega = |u| \cos \theta$; special cases in $d = 3$ dimensions are

- $\kappa = -3$ for Coulomb potential,
- $\kappa = 0$ for Maxwell potential $U(q) = C|q|^{-4}$,
- $\kappa = 1$ for hard spheres (see (12)),

with appropriate functions β.

The local density n, average velocity V and temperature T (if $n \neq 0$) are

$$n(q, t) = \int_{\mathbb{R}^d} f(q, v, t) \, dv \tag{30}$$

$$V(q, t) = \frac{1}{n} \int_{\mathbb{R}^d} v f(q, v, t) \, dv \tag{31}$$

$$T(q, t) = \frac{1}{nd} \int_{\mathbb{R}^d} |v|^2 f(q, v, t) \, dv \tag{32}$$

---

[4] The unit vector ω spans the sphere S^{d-1} of unit radius in $\mathbb{R}^d$.

Define the local instantaneous rate of entropy production

$$\mathcal{D}_f(q,t) = \frac{1}{4} \int_{\mathbb{R}^d} \int_{\mathbb{R}^d} \int_{S^{d-1}} B(v - v_*, \omega)(f'f'_* - ff_*) \ln \frac{f'f'_*}{ff_*} d\omega dv_* dv \qquad (33)$$

and Boltzmann's local $\mathcal{H}$ function

$$\mathcal{H}_f(q,t) = \int_{\mathbb{R}^d} f(q,v,t) \ln f(q,v,t) dv \qquad (34)$$

Given a distribution function $M(v) > 0$ (normalized by $\int_{\mathbb{R}^d} M dv = 1$), we define (minus) the relative entropy of f with respect to M as

$$\mathcal{H}_{f|M}(q,t) = \int_{\mathbb{R}^d} f(q,v,t) \ln \frac{f(q,v,t)}{n(q,t)M(v)} dv \qquad (35)$$

One shows that $\mathcal{H}_{f|M} \geq 0$, and that $\mathcal{H}_{f|M} = 0$ only if $f = nM$.

The following theorems [6,7] apply to the spatially homogeneous Boltzmann equation and are being extended to more complex cases.

Theorem: *Let the exponent $0 < \kappa \leq 1$ in the collision kernel (29). If $\int_{S^{d-1}} \beta(r \cdot \omega) d\omega < \infty$ for any $r \in S^{d-1}$, and given the initial distribution function $f(q,v,0) = f_0(v)$, with density n_0, average velocity $V_0 = 0$ and temperature T_0, let M be the corresponding maxwellian distribution, $M = M_{T_0}$. Then*

1. *The Boltzmann equation (28) with initial condition f_0 has a unique solution for all time $t \geq 0$, and this solution f conserves mass (mn_0), momentum $(mV_0 = 0)$ and energy (T_0). The solution f also satisfies*

$$\frac{\partial}{\partial t} \mathcal{H}_{f|M} = -\mathcal{D}_f \qquad (36)$$

2. *For any $t_0 > 0$, there exist constants $K > 0$, $A > 0$ such that $\forall t > t_0$: $f(v,t) \geq Ke^{-A|v|^2}$.*
3. *There exist $c > 0$, $C > 0$ such that $\int_{\mathbb{R}^d} |f(v,t) - n_0 M(v)| dv \leq Ce^{-ct}$.*

This theorem ensures that the initial (spatially homogeneous) distribution function will relax to the maxwellian distribution M having the same invariants, that the tails of the distribution cannot decay faster than a maxwellian, and that the approach to the asymptotic maxwellian is exponential.

Furthermore, the evolution to equilibrium is also controlled [7]:

Theorem: *If $B \geq 1$ and if there exist $K > 0$, $A > 0$ such that $f_0(v) \geq Ke^{-A|v|^2}$, with $V_0 = 0$, then for any $\epsilon > 1$ there exists $C' > 0$ such that the entropy production rate satisfies $\mathcal{D}_f \geq C'(\mathcal{H}_{f|M})^\epsilon$.*

This additional statement shows that the entropy production rate cannot be small (i.e. the relaxation cannot proceed arbitrarily slowly) if the distribution function is far from the equilibrium maxwellian – provided that the differential cross sections remain large for all u and ω (i.e. ensure a 'fast' redistribution of velocities). The condition $B \geq 1$ can be relaxed to deal with $\kappa > 0$ (then $\epsilon > 0$).

Acknowledgements

It is a pleasure to thank C. Froeschlé, D. Benest and P. Michel for organizing this school and CNRS for supporting it. I am grateful to lecturers – especially A. Celletti, C. Falcolini, G. della Penna and J. Waldvogel – for their teaching, and to participants for fruitful and pleasant discussions. Finally, I am indebted to M. K-H. Kiessling for Glassey and Schaeffer's results.

References

1. W. Appel and M. K-H. Kiessling: 'Mass and spin renormalization in Lorentz electrodynamics', *Ann. Phys. (NY)* **289** (2001) 24-83
2. R. Balescu: *Statistical dynamics – Matter out of equilibrium* (Imperial college press, London, 1997)
3. J.R. Dorfman: *An introduction to chaos in nonequilibrium statistical mechanics* (Cambridge university press, 1999)
4. R. Glassey and J. Schaeffer: 'On symmetric solutions of the relativistic Vlaso-Poisson system', *Comm. Math. Phys.* **101** (1985) 459-473
5. H. Spohn: *Large scale dynamics of interacting particles* (Springer, New York 1991)
6. C. Villani: Contribution à l'étude mathématique des équations de Boltzmann et de Landau en théorie cinétique des gaz et des plasmas. Thèse de doctorat, CERE-MADE, université Paris IX-Dauphine (1998)
7. C. Villani: 'On the trend to equilibrium for solutions of the Boltzmann equation: quantitative versions of Boltzmann's *H*-theorem', preprint, Ecole normale supérieure, Paris, 1999

Chaotic Scattering in Planetary Rings

Jean-Marc Petit

Observatoire de la Côte d'Azur, Observatoire de Nice, B.P. 4229
F-06304 Nice Cedex 4, France

Abstract. The gravitational interaction of two small satellites, on initially close circular, coplanar orbits leads to a one-parameter family of solutions. When varying the parameter h of the family, the solution changes continuously on an interval and then undergoes a sudden change. The set of discontinuities has a Cantor-like structure.

Similar phenomena have been observed in other problems of scattering. Such a behavior is related to the presence of periodic orbits and homo- and heteroclinic points. It can be shown that in the vicinity of a homo- or heteroclinic point, one can define a symbolic dynamics (Moser, 1973).

The large eigenvalue (600) of the satellite problem limits the possibility of numerical exploration. A model problem, the inclined billiard (Hénon, 1988), was designed with a tunable eigenvalue, for which the symbolic dynamics can be analytically defined, thus fully elucidating the structure of the family.

1 Introduction

In the last few years, many studies have been carried out on the chaos in bound classical hamiltonian systems and powerful methods have been developed and applied. In contrast, less work has been done on chaos in classical scattering systems. However, for nearly thirty years, there have been numerical observations of complicated - chaotic - behavior in continuous scattering problems: classical models for inelastic molecular scattering (Rankin and Miller 1971, Gottdiener 1975, Fitz and Brumer 1979, Schlier 1983, Noid et al. 1986), satellite encounters (Petit and Hénon 1986), vortex dynamics (Eckhardt and Aref 1989), potential scattering (Eckhardt and Jung 1986, Jung and Scholz 1987). But until recently, this phenomenon had not been studied for itself.

We found this kind of behavior in a simple physical problem: the encounter of two satellites on close circular orbits around a planet (namely, Saturn). In section 2, we describe in detail the physical problem and derive the equations of motion. One would notice that the equations of motion are non integrable and contain no true singularities. In the same section, we present more precisely the chaos that appears in our problem: the asymptotic behavior of the system is discontinuous with respect to the initial parameters. Then, in section 3, we give some theoretical results on symbolic dynamics and the generality of this kind of chaotic behaviour. In view of the numerical difficulties which were encountered in exploring this problem, a simple "model" problem was developped (Hénon 1988) which is just complex enough to exhibit all the features we are interested in. This model is described in section 4, and we present the derivation of the associated symbolic dynamics.

2   Dynamics of planetary rings

2.1   The physical problem

The physical system we consider consists in the planetary rings, more the Saturn rings. They represent extremely flat structures, with a length of $\sim 6 \cdot 10^5$ km, a width of $\sim 7 \cdot 10^4$ km and a thickness of ~ 100 m. Except for some particular narrow rings, they are almost circular. We are interested in rings of moderate optical thickness. Thus 2-body interactions are important, but we can neglect n-body interactions (n ≥ 3).

Hence the model we consider is a particular case of the three body problem (Fig. 1). Two light bodies M_2 and M_3 describe initially coplanar and circular orbits, with slightly different radii (a_2 and a_3, with $|a_3 - a_2| = \Delta a \ll (a_2, a_3)$), around a heavy central body M_1. Bodies M_2 and M_3 are initially far apart, so that their mutual attraction is negligible. However, the inner body has a slightly larger angular velocity and eventually catches up with the outer body; the distance from M_2 to M_3 becomes small and their mutual attraction is no longer negligible. We shall call this an *encounter*. For convenience, M_1 will be called the *planet* and M_2, M_3 will be called the *satellites*. The difference between the radii of the initial circular orbits Δa will be called the *impact parameter*.

Analytic approximations of the solution are available in two cases:

(i) When Δa is sufficiently large, the result of the encounter is only a slight deflection of M_2 and M_3 from their previous circular orbits (Fig. 2). These deflections can then be obtained by a perturbation theory (Goldreich and Tremaine 1979, 1980).

(ii) When the impact parameter is very small, the interaction of M_2 and M_3 produces a "horseshoe" motion: M_2 and M_3 "repel" each other azimutally and never come in close proximity (Fig. 3). This case can also be treated by a perturbation theory (Dermott and Murray 1981, Yoder et al. 1983). It corresponds to an adiabatic invariant (Henon and Petit 1986).

Between these two asymptotic cases, however, no theory exists, and apparently only a numerical integration of the equations of motion can give the answer.

2.2   Equations of motion

The equations of motion are as follows:

$$\ddot{X}_i = \frac{Gm_j(X_j - X_i)}{R_{ij}^3} + \frac{Gm_k(X_k - X_i)}{R_{ik}^3},$$

$$\ddot{Y}_i = \frac{Gm_j(Y_j - Y_i)}{R_{ij}^3} + \frac{Gm_k(Y_k - Y_i)}{R_{ik}^3},$$

where for $i = 1, 2, 3$ and $i \neq j \neq k$. R_{ij} is the distance between body i and body j. These equations are not well suited for accurate numerical integration. One has to compute the difference of large, very close numbers, hence loosing a lot of significant digits.

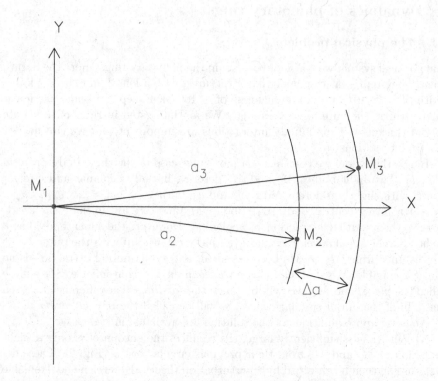

Fig. 1. Plane view of the three body problem defining the notations.

In order to have an accurate numerical study, we first reduce the equations to a simpler form: the classical set of Hill's equations. Only a brief review of this reduction will be given here; details can be found in Hénon and Petit 1986. We call m_i is the mass of body M_i and $m = m_1 + m_2 + m_3$ the total mass of the system.

- We assume that the mass of either satellite is small compared to the mass of the planet:

$$m_2 \ll m_1, \qquad m_3 \ll m_1. \tag{1}$$

- We assume also that the distance between the two satellites is small compared to their distance to the planet. In a zero-order approximation, the two satellites can then be considered as a single body in orbit around the planet. This orbit will be called the *mean orbit*, and will be assumed to be circular.
  - We call a_0 the radius of the mean orbit. (The precise definition of a_0 does not matter, as long as it is nearly equal to the radii of the satellite orbits).
  - The angular velocity on the mean orbit is

$$\omega_0 = \sqrt{Gma_0^{-3}}. \tag{2}$$

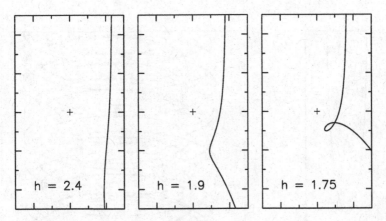

Fig. 2. Some slightly perturbed orbits for large values of Δa. The curve represents the relative motion of one satellite with respect to the other, in Hill's coordinates (ξ in abscissa, η in ordinate; see text below for definitions). The first approach is downwards from $\eta = +\infty$.

We define the relative mass of this single body

$$\mu = \frac{m_2 + m_3}{m}. \tag{3}$$

Let X_i, Y_i be the coordinates of body i in an inertial system and t the time. We introduce dimensionless coordinates by

$$X_i' = \frac{X_i}{a_0}, \qquad Y_i' = \frac{Y_i}{a_0}, \qquad m_i' = \frac{m_i}{m}, \qquad t' = \omega_0 t, \tag{4}$$

and for simplicity we drop the primes in what follows.

- In the new variables, the radius of the orbit, the angular velocity, the mass of the system, and the gravitational constant are all equal to 1.
- We choose the origin of time so that the two satellites are in the vicinity of $X = 1$, $Y = 0$ at $t = 0$.
- In a system rotating at angular velocity 1, the two satellites remain close to point $(1, 0)$ during the encounter.
- We introduce new coordinates ξ, η, which will be called *Hill's coordinates*:

$$X_i - X_1 = (1 + \mu^{1/3}\xi_i) \cos t - \mu^{1/3}\eta_i \sin t,$$
$$Y_i - Y_1 = (1 + \mu^{1/3}\xi_i) \sin t - \mu^{1/3}\eta_i \cos t, \qquad (i = 2, 3) \tag{5}$$

In the previous derivation, we introduced the small parameters $\mu^{1/3}\xi$ and $\mu^{1/3}\eta$. We scaled by the factor $\mu^{1/3}$ because we want to have a non-trivial problem. If the two satellites are far apart, then we have the superposition of two 2-body Keplerian problems. If on the contrary the two satellites are very close, then we have a 2-body problem, which center of mass is on a Keplerian orbit

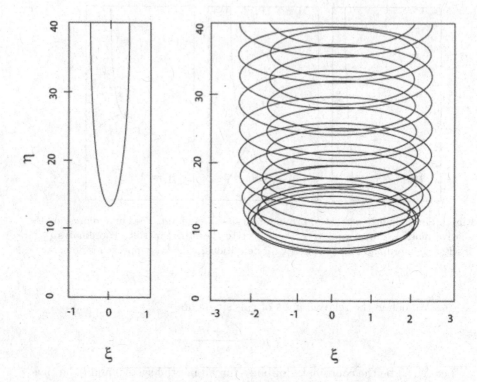

Fig. 3. Same as Fig. 2 but for small Δa.

around the massive body. But if the gravitational force between the two satellites is similar to the differential force due to the central body, then we cannot neglect any force term in the equations of motion:

- the distance to the primary is $O(1)$;
- the differential force is then $O(R_{23})$;
- the gravitational force is $O(\mu/R_{23}^2)$;
- we finally have the relation $O(R_{23}) \sim O(\mu/R_{23}^2)$.

Hence:
$$R_{23} = O(\mu^{1/3}),$$
from where we obtain the small parameter.

We now consider the motion of the center of mass and the relative motion of the two satellites. In all what follows, we drop the terms of order $\mu^{1/3}$ or higher.

The position of the center of mass is then:
$$\xi^* = \frac{m_2\xi_2 + m_3\xi_3}{m_2 + m_3}, \qquad \eta^* = \frac{m_2\eta_2 + m_3\eta_3}{m_2 + m_3},$$
and it satisfies (approximately) the equations of motion:
$$\ddot{\xi}^* = 2\dot{\eta}^* + 3\xi^*, \qquad \ddot{\eta}^* = -2\dot{\xi}^*.$$

These equations are linear and easily integrated as an epicyclic motion (Hénon and Petit 1986):

$$xi^* = D_1^* \cos t + D_2^* \sin t + D_3^*,$$
$$\eta^* = -2D_1^* \sin t + 2D_2^* \cos t - \frac{3}{2} D_3^* t + D_4^*.$$

The relative position is:

$$\xi = \xi_3 - \xi_2, \qquad \eta = \eta_3 - \eta_2,$$

satisfying (approximately) the *Hill's equations* (Hill 1978):

$$\ddot{\xi} = 2\dot{\eta} + 3\xi - \frac{\xi}{\rho^3}, \qquad \ddot{\eta} = -2\dot{\xi} - \frac{\eta}{\rho^3}, \qquad \rho = \sqrt{\xi^2 + \eta^2}.$$

The error in these equations is of order of $\mu^{1/3}$. They become exact in the limit of vanishing satellite masses. Taking this limit is equivalent to zoom on the two satellites and the main effect is to repel the planet to infinity and transform circular orbits in straight lines (fig 4).

The most important points to notice on these equations are:

1. There is no parameter left in the equations (the same equations are valid in every physical case).

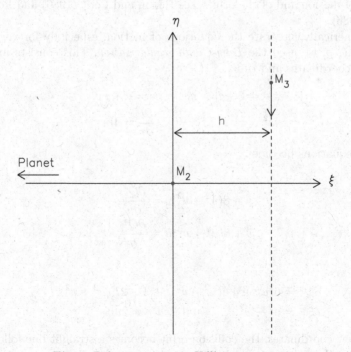

Fig. 4. Relative motion in Hill's approximation.

2. As is easily shown, the initial conditions for relative motion of circular orbits are given by only one parameter: the impact parameter h.
3. Therefore, the set of solutions is a one-parameter family, and it seems reasonable to try to study it. We can even reduce the study to positive values of h because of the symmetries of the equations of motion.

Hill's equations admit the integral

$$\Gamma = 3\eta^2 + \frac{2}{\rho} - \dot{\xi}^2 - \dot{\eta}^2 \tag{6}$$

which can be called the *Jacobi integral* by analogy with the restricted problem. We can write the Jacobi integral in terms of the initial conditions:

$$\Gamma = \frac{3}{4}h^2. \tag{7}$$

The typical encounter orbits are as shown in fig 5, taken from a collection of several hundred pictures of the familly $0 < h < \infty$.

For convenience, we shall think of the special case $m_2 \gg m_3$, and identify the origin of the (ξ, η) system with the satellite M_2; the curves then simply represent the motion of M_3. An interesting feature is that the third body always escapes either upward or downward, but never stays close for ever. This is in agreement with a general result by Marchal (1977) which shows that the set of "capture orbits" is of zero measure. For a more detailed explanation of the equation of motion and of the orbits, see Hénon and Petit (1986) and Petit and Hénon (1986).

To numerically integrate the equations of motion, especially for very small or vanishing ρ, we used the *Levy-Civita regularization*. This consists in a new change of coordinates and time:

$$z = \xi + i| > \eta, \qquad w = u + iv,$$
$$w^2 = z, \qquad \frac{\partial t}{\partial \tau} = 4|w|^2$$

Then the equations become:

$$\ddot{u} - 8(u^2 + v^2)\dot{v} = \frac{\partial Q^*}{\partial u},$$
$$\ddot{v} + 8(u^2 + v^2)\dot{u} = \frac{\partial Q^*}{\partial v},$$

with

$$Q^* = 6(u^2 + v^2)(u^2 - v^2)^2 + 4 - 2\Gamma(u^2 + v^2),$$
$$\xi = u^2 - v^2, \qquad \text{and} \qquad \eta = 2uv.$$

In these new coordinates, the collision orbit becomes a straight line followed at constant speed.

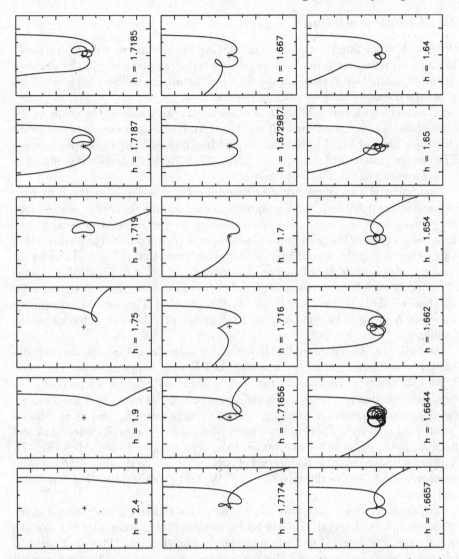

Fig. 5. Family of encounter orbits. Each frame corresponds to one particular value of the reduced impact parameter h. The curve represents the relative motion of one satellite with respect to the other, in Hill's coordinate (ξ in abscissa, η in ordinate). The first approach is downwards from $\eta = +\infty$.

It is interesting to note the following propoerties:

1. there is no true singularity in the equations of motion: the $1/\rho^2$ singularity can be removed by the Levy-Civita regularization;
2. at a given time t, the solution is a continuous function of the impact parameter h;
3. the orbits are *not* chaotic.

2.3   Chaotic scattering

When $t \to \infty$ the family exhibits an interesting feature that we call "transitions". Roughly speaking, at given values of h, we see a discontinuity in the shape of the orbit: orbits that used to escape with $\eta < 0$ starts to escape with $\eta > 0$. This is the phenomenon that we want to develop now.

Consider an example. When h decreases from large values, the shape of the orbit changes continuously with h and the third body always escapes downward (first four plots of Fig. 5). Suddenly, something happens and it escapes upward. The change (transition) occurs for $h_{max} = 1.718779940$. It can be thought of as a discontinuity of the shape of the orbit.

But we need a more quantitative description of this discontinuity. If we look at parameters describing the asymptotic motion as functions of h, we see very sharp variations at values of h corresponding to the changes of escape side. Especially, consider the final impact parameter h' (defined from the mean motion for $t \to \infty$). Using the Jacobi integral, it can be shown that $|h'| \geq h$. A downward escape corresponds to $h' > 0$ and an upward escape to $h' < 0$. Therefore a change of escape side leads to a change of sign for h' and a discontinuity of step at least $2h$ (Fig. 6). This is what we mean by discontinuity. The set of discontinuity values of h being very complex, we shall speak of "chaotic" behavior of the familly.

We will now describe rapidly the set of discontinuities. Consider an orbit defined by an arbitrary value h_0. Typically, the following happens: when decreasing h from h_0, the orbit changes continuously down to h_1 where a discontinuity occurs. We call this a "transition value". Similarly, if we increase h from h_0, we reach a second transition value h_2. The interval between h_1 and h_2 is called a "continuity interval". There are two particular cases: a continuity interval ranges from h_{max} to ∞, an other one ranges from 0 to $h_{min} = 1.336117188$ (Fig. 7). Suppose we have localized an interval of continuity. We do it again, starting from another value h_0 out of the range $[h_1, h_2]$. We find another interval of continuity and so on.

Experiment shows that intervals are never contiguous. If one takes a point in an unexplored interval, one will find a new continuity interval which doesn't touch a previous interval neither on the left nor on the right. This gives birth to two new unexplored intervals. This goes on and on to infinity. This must remind the reader of the classical definition of the Cantor set. The difference here is that the intervals are not regularly ordered.

In order to reconcile the continuity of the orbits with the discontinuity of the asymptotic behavior (h'), the family must go through an orbit with infinite capture time. This is achieved by having an orbit asymptotic to a periodic orbit (Fig. 8). For example, the first transition orbit for $h = h_{max}$ corresponds to a periodic orbit emanating from the Lagrangian point L_2 (Hénon 1969, Fig. 2). Since the system is Hamiltonian, the periodic orbit is necessarily unstable.

It will be helpful to introduce at this point a *surface of section* defined for instance by $\eta = 0$ and $\dot{\eta} > 0$: for each crossing of an orbit with the ξ axis in

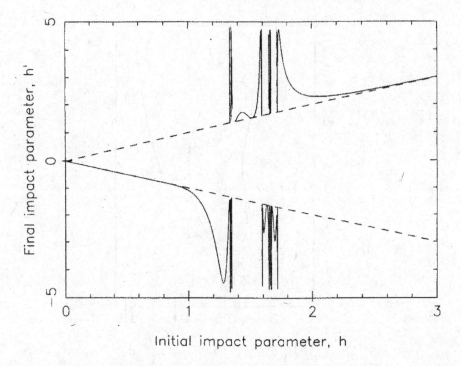

Fig. 6. Final impact parameter h' as a function of the initial impact parameter h. The region between the two dashed lines is forbidden.

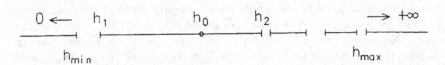

Fig. 7. A Schematic representation of the largest continuity intervals.

the positive direction (η increasing), we plot a point with the coordinates ξ, $\dot{\xi}$ (Fig. 9). An orbit is then represented by a sequence of points.

On the figure, P represents the periodic orbit. The two real eigenvalues of this orbit are $\lambda_1 = 1/640$ and $\lambda_2 = 640$. The orbit $h = h_{max}$ falls on the stable invariant manifold W_s of P and converges exponentially to P: points $(\cdots, Y_0, Y_1, Y_2, \cdots)$. There are orbits tending towards the periodic orbit for $t \to -\infty$: points $(\cdots, Z_{-2}, Z_{-1}, Z_0, \cdots)$ on the unstable invariant manifold W_u of P. Note however that for the orbits we are concerned with, we get generally a finite (small) number of points in the surface of section: three points for the orbit with

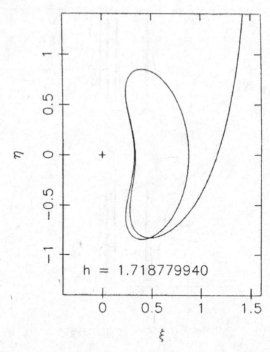

Fig. 8. An orbit of the Satellite Encounter family which is asymptotic to an unstable periodic orbit.

$h = 1.71863$ for instance. An orbit can also have no point at all in that surface ($h > 2.4$).

Consider now an orbit of our family with h slightly different from h_{max}, say larger. The points in the surface of section are slightly beside W_s (crosses on the picture). They stay close to W_s until they reach the vicinity of P, then they go away along W_u. An important point is that λ_2 is positive. So the points go along only one branch of W_u. Here, it is the upper right branch. The corresponding orbits are quite regular. Particularly, they all escape downward and vary continuously when h increases (Fig. 10a). This accounts for the continuity interval for $h > h_{max}$.

For $h < h_{max}$, the points escape along the left branch of W_u. The orbits for $h < h_{max}$ are shown on Fig. 10b. Sometimes orbits escape upward, some time downward. So there is no continuity interval on the left of h_{max}. This explains the complex structure of the continuity intervals. For $h < h_{max}$, instead of escaping directly, the orbit will first go in the vicinity of an other unstable periodic orbit. This orbit will itself give birth to a transition phenomenon, that we shall call a *second order transition*. In this way, one can construct a hierarchical structure of transitions of higher and higher order. Suppose we have an orbit going close to one periodic orbit then close to a second one. By changing h, we can push the points in the surface of section closer to the first fixed point. Particularly,

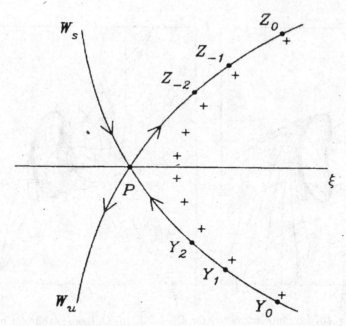

Fig. 9. Sketch of the surface of section. The value of λ_1 has been artificially increased to show the structure more clearly.

one can manage to have the same pattern along W_u and one or more additional points in the vicinity of P. This corresponds to orbits with the same escape but with one or more additional turns around the first periodic orbit (Fig. 11). In the first plot, the orbit follows the periodic orbit during half a turn, in the second during one and a half and in the third during two and a half (even if this is not visible on the figure). This gives rise to a geometrical progression of ratio λ_1 in the values of h.

From all our numerical integrations, it seems that only two family of periodic orbits are involved: family a mentioned above and the symmetrical family b also described in Hénon 1969.

The necessary ingredients for this kind of behavior is the existence of periodic orbits and heteroclinic or homoclinic points (intersection points of invariant manifolds of two different or one single periodic orbit). But it is very difficult to go any further with this problem due to the large value of the eigenvalue (~ 640).

2.4 Other examples

Other authors have observed similar behavior in other scattering problems: collisions of vortex pairs (Manakov and Shchur, 1982; Eckhardt and Aref, 1989; Fig. 12), collisions between an atom and a diatom (Gottdiener, 1975; Agmon, 1982; Fig. 13), scattering of a particle by a two-dimensional potential (Jung and

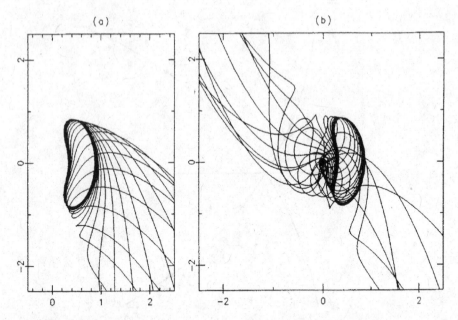

Fig. 10. (a) Outgoing orbits for $h > h_{max}$. (b) Outgoing orbits for $h < h_{max}$.

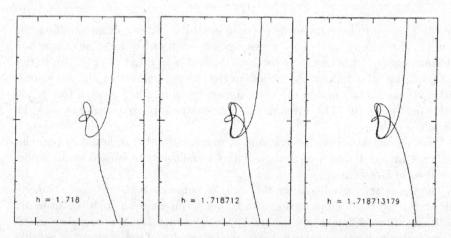

$h = 1.718$ $h = 1.718712$ $h = 1.718713179$

Fig. 11. Three orbits with essentially the same outgoing but different behavior during the close encounter. The orbit goes along the unstable periodic orbit for half a turn on the left, for one and a half in the middle and for two and a half on the right plot.

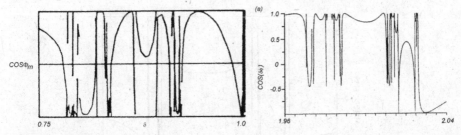

Fig. 12. Collisions of vortex pairs. Top: Manakov and Shchur, 1982. Bottom: Eckhardt and Aref, 1987.

Scholz, 1987; Eckhardt and Jung, 1986; Blümel and Smilansky, 1987; Fig. 14) among others.

3   Symbolic dynamics

In this section, we present very briefly some general results concerning the symbolic dynamics and the effect of homo- and heteroclinic points. In particular,

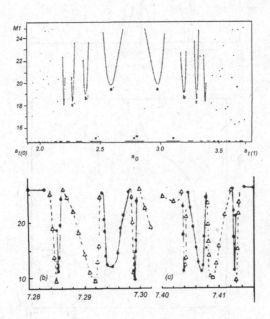

Fig. 13. Collisions between an atom and a diatom. Top: Gottdiener, 1975. Bottom: Agmon, 1982.

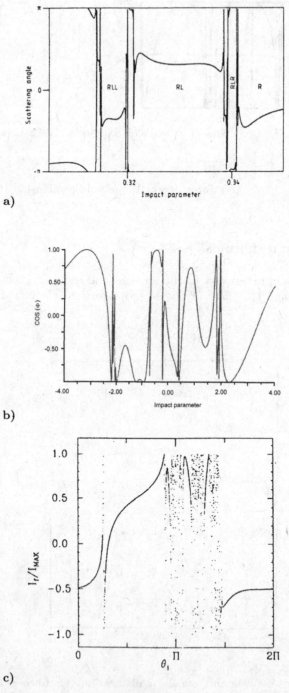

Fig. 14. a) Diffusion by a two-dimensional potential. Jung and Scholz, 1987. **b)** Jung and Scholz, 1987. **c)** Blümel and Smilansky, 1987.

no proof will be given. We refer the readers interested in more mathematical derivations and proofs to Moser (1973) and references therein.

We consider the topological space S consisting in doubly infinite sequences

$$s = (\cdots, s_{-2}, s_{-1}, s_0, s_1, s_2, \cdots)$$

with $s_k \in A$, a finite or denumerable set which we call *the alphabet*.

The goal of *symbolic dynamics* is to relate the orbits of a mapping or of a dynamical system with the elements of S.

3.1 The Bernoulli shift

On S, we define the mapping σ such that:

$$(\sigma(s))_k = s_{k-1}.$$

This mapping is known as the *Bernoulli shift*.

We consider now the mapping ϕ on the square Q: $0 \le x \le 1$, $0 \le y \le 1$ (Fig. 15):

$$\phi : \begin{cases} x \to x_1 = 2x - [2x], \\ y \to y_1 = \frac{1}{2}(y + [2x]). \end{cases}$$

This mapping is known as the *Baker's transform*.

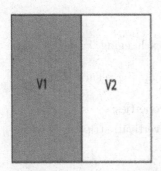 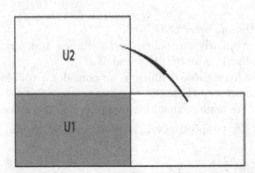

Fig. 15. Geometrical definition of the Baker's transform: the square Q is first stretched by a factor 2 along the horizontal axis, while compressed by the same factor along the vertical axis. Then the right half is cut and put atop the left half.

We now want to relate the elements of S to the elements of Q in such a way that the mapping σ on S corresponds to the mapping ϕ on Q. Taking $A = 0, 1$, we define a mapping τ from S to Q by:

$$x = \sum_{k=0}^{-\infty} s_k 2^{k-1}, \qquad y = \sum_{k=1}^{+\infty} s_k 2^{-k}.$$

One can easily see that we have $\phi = \tau\sigma\tau^{-1}$. Geometrically, the two vertical strips V_1 and V_2 of Fig. 15 are described by $s_0 = 0$ and 1. They are mapped into the two horizontal strips U_1 and U_2, described by $s_1 = 0$ and 1, thus corresponding to a shift in the sequence s.

Therefore, we can specify arbitrarily if the images and preimages of a point belong to V_1 or V_2. σ is then called a subsystem of ϕ.

3.2   Topological mappings

The properties we just described for the Baker's transform can be shown to apply to more general tranforms.

We first need to introduce some definitions. We define, in Q, a *horizontal line* $y = u(x)$ by:

$$0 \leq u(x) \leq 1 \qquad \text{for} \qquad 0 \leq x \leq 1,$$

and

$$|u(x_1) - u(x_2)| \leq \mu|x_1 - x_2| \qquad \text{for} \qquad 0 \leq x_1 \leq x_2 \leq 1$$

with $0 < \mu < 1$.

Let $u_1(x)$ and $u_2(x)$ be two horizontal lines such that $0 \leq u_1(x) < u_2(x) \leq 1$. We call the set

$$U = \{(x,y) \mid 0 \leq x \leq 1;\ u_1(x) \leq y \leq u_2(x)\}$$

a *horizontal strip*.

The maximum distance along the vertical axis between the two horizontal lines,

$$d(U) = \max_{0 \leq x \leq 1} (u_2(x) - u_1(x))$$

is the *diameter* of U.

Similarly we define *vertival* lines and strips by exchanging x and y, and replacing u and U by v and V.

Given these definition, we consider a topological mapping ϕ such that $\phi(Q)$ intersects Q (Fig. 16).

We assume that this mapping has the following properties:

(1) U_a (respectively V_a) are disjoint horizontal (resp. vertival) strips in Q with:

$$\phi(V_a) = U_a, \qquad a \in A.$$

The vertical boundaries of V_a are mapped onto the vertical boundaries of U_a.

(2) If $V \in \cap_{a \in A} V_a$, then for any $a \in A$

$$\phi^{-1}(V) \cap V_a = V'_a$$

is a vertical strip, and for $0 < \nu < 1$, we require

$$d(V'_a) \leq \nu d(V_a).$$

It can be shown that if ϕ satisfies the conditions (1) and (2), then it possesses σ as a subsystem: there exist τ from S into Q such that

$$\phi\tau = \tau\sigma.$$

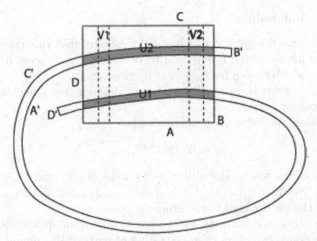

Fig. 16. Sketch of a general topological mapping ϕ such that $\phi(Q)$ intersects Q twice. The two vertical strips V_1 and V_2 are mapped by ϕ into the horizontal strips U_1 and U_2 respectively.

3.3 C^1 mappings

We consider now a mapping ϕ which is C^1:

$$\phi : \begin{cases} x_1 = f(x_0, y_0), \\ y_1 = g(x_0, y_0), \end{cases}$$

where f and g are two C^1 functions on Q.

We call $d\phi$ its tangent mapping:

$$d\phi : \begin{cases} \xi_1 = f_x \xi_0 + f_y \eta_0, \\ \eta_1 = g_x \xi_0 + g_y \eta_0, \end{cases}$$

where f_x, f_y, g_x and g_y denote the partial derivatives of f and g with respect to x and y.

Let us define the following condition.

(3) For $0 < \mu < 1$, we define the bundle of sectors

$$S^+ : |\eta| \leq \mu|\xi|$$

over $\cup_{a \in A} V_a$. We assume that

$$d\phi(S^+) \subset S^+.$$

In addition, if $(\xi_0, \eta_0) \in S^+$ and (ξ_1, η_1) is its image, then

$$|\xi_1| \geq \mu^{-1}|\xi_0|.$$

Similarly we define the bundle of sectors S^- over $\cup_{a \in A} U_a$, and have the same properties changing ξ and η.

It can be shown that if ϕ satisfies the conditions (1) and (3) with $0 < \mu < 1/2$, then condition (2) holds with $\nu = \mu/(1 - \mu)$.

3.4   Homoclinic points

We now consider a mapping ϕ which is C^∞. Assume that this mapping has 2 different fixed points, p and q. r is said to be a *heteroclinic point* if $\phi^k(r) \to p$ for $k \to -\infty$ and $\phi^k(r) \to q$ for $k \to +\infty$, or vice-versa.

In case the 2 fixed points p and q are not different but coincide, then r is called a homoclinic point.

In both cases, the following property holds:

$$r \in W_u(p) \cup W_s(q).$$

In what follows, we require that the 2 curves $\dot{W}_u(p)$ and $W_s(q)$ intersect transversally. We restrict ourself to the case of homoclinic points, i.e. p and q coincide, but the results hold for heteroclinic points.

Given the previous conditions, we can construct a small quadrilateral R near r, homoclinic point, two of its sides consisting of parts of $W_u(p)$ and $W_s(p)$ and the others being straight line parallel to the tangents of $W_u(p)$ and $W_s(p)$ at r (Fig. 17).

For any given point q, we call $k = k(q)$ the smallest integer for which $\phi^k(q) \in R$.

We call $D(\phi')$ the set of $q \in R$ for which $k > 0$, and

$$\phi'(q) = \phi^k(q) \qquad \text{for} \qquad q \in R.$$

The following results can be shown:

- In any neighborhood of r, ϕ' possesses an invariant subset I homeomorphic to S via $\tau : S \to I$ such that

$$\phi'\tau = \tau\sigma.$$

- The homoclinic point belonging to p are dense in I.

As a consequence, for a dynamical system with two periodic orbits and a heteroclinic point or a single periodic orbit with a homoclinic point, there would be an unnumerable set of initial conditions for which one could define a symbolic dynamics.

4   The inclined billiard

According to the previous results, it is theoretically possible to define a symbolic dynamics which is Bernoulli for the three body problem. This gives a better description of the dynamics of the system. But in our problem, it is difficult to study the symbolic dynamics because of the large value of the eigenvalues. This greatly limits the number of scales that we can explore. Hence it is not possible to explicit the self-similar structure of the values of discontinuity.

So a model problem was designed which is complex enough to exhibit all the features we are interested in, and simple enough so that all the calculations can be done analytically. This new model should satisfy the following conditions:

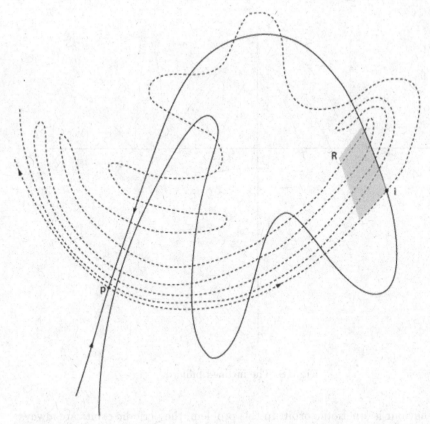

Fig. 17. Schematic representation of the stable and unstable manifolds $W_u(p)$ and $W_s(q)$ of a fixed point p, a homoclinic point r, and a small quadrilateral R in its neighborhood.

- have smaller, and if possible adjustable eigenvalues;
- the equations of motion can be reduced to an explicitly defined mapping, so that solutions can be computed much faster.

4.1 The model and an interesting limit

This model is the *inclined billiard* (Hénon, 1988). It is defined as follows: a particle moves in the (X, Y) plane and bounces elastically on two fixed disks with radius r and with their centers in $(-1, -r)$ and $(1, -r)$ respectively. In addition, it is subjected to a constant acceleration g which pulls it in the negative Y direction (Fig. 18).

For most initial conditions, after a finite number of rebounds, the particle will "fall" downwards and never return. This behaviour is equivalent to a departure phase in Hill's problem. However, for some particular conditions, the particle can bounce indefinitely back and forth on one or both disks. A particular instance of

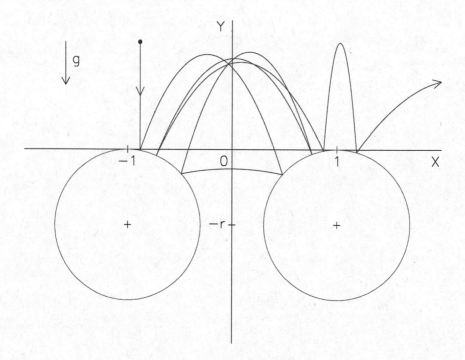

Fig. 18. The inclined billiard

this behaviour is a periodic orbit. In this problem, the periodic orbits are always unstable.

To follow the same reasoning path as for Hill's problem, we define a one-parameter family. We start the particle at rest at a position given (h, Y_0). Y_0 is the constant height from which we drop the particle, h is the variable location along the horizontal axis, similar to the impact parameter in Hill's problem.

We describe the family by varying h from $-\infty$ to $+\infty$. The setup of the system is such that we expect intervals of continuity in h. We also expect transitions:

- For $h = -1$, the particle bounces indefinitely on the left disk: this is a periodic orbit.
- For $h < -1$, the particle falls to the left and never returns.
- For $h > -1$, it moves to the right and complex interplays with the 2 disks are possible.
- A similar periodic orbit exists at $h = 1$.

The particle being subjected to a constant acceleration, the orbits are piecewise parabolae. Hence the problem can in principle be reduced to the study of an explicit mapping. To do so, we consider the coordinates at rebound. The coordinate of the rebound X and Y are linked by the fact that the rebound takes place at the surface of a disk. We keep X as one of our mapping coordinate. The energy of the particle is fixed and given by the initial conditions. So, another coordinate can be dropped. We decide to keep W, the transverse component of the velocity, i.e. the projection of the velocity on the tangent to the disk. W is invariant in the rebound. So we do not have to specify if it is taken before or after the rebound. This choice preserves the symmetry of the mapping with respect to time. Eventhough theoretically possible, the exact calculation of the mapping is messy. it requires to solve a 4th order equation.

To make the computation affordable, one considers the limit where r is large and approximates the circles (disks) by parabolas (Fig. 19). The "disks" extend then from $-\infty$ to ∞ in the X direction and the number of rebounds of the particle on them is now always infinite. We suppose that Y_0 is large.

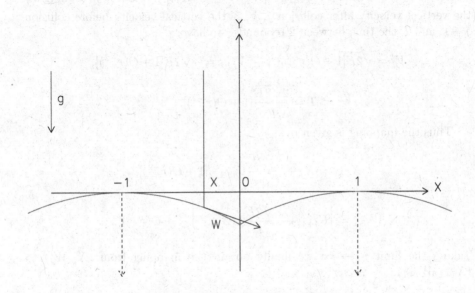

Fig. 19. The limit of large radii.

In this limit, we can make many simplifying approximations. Y (at the collision point) is small: $Y = O(r^{-1})$. The vertical velocity V is large. So we can neglect the vertical thickness of the profile. Only the slope $f(X)/r$ is of consequence:

$$f(X) = \begin{cases} X+1 & \text{for } X < 0, \\ X-1 & \text{for } X \geq 0, \end{cases}$$

The constant total energy is $E = gY + \frac{1}{2}(U^2+V^2)$, where $U = \dot{X}$ and $V = \dot{Y}$. The change in $U/V = O(1/r)$, thus, for an interesting mapping, we consider the case where $U/V = O(1/r)$. The time between rebounds is $O(V/g)$, ther horizontal distance travelled is $O(UV/g)$. Here again, for an interesting mapping, this is

$O(1)$. Hence:

$$UV = O(g), \quad U = O(\sqrt{\tfrac{g}{r}}), \quad V = O(\sqrt{gr}), \quad E = O(gr).$$

The maximum height reached is $Y_{max} = O(r)$.

We then derive the mapping. We first introduce some notations. X_j and Y_j are the coordinates of collision number j. R_j and W_j are the radial and tangential velocity components just after the collision. We define θ as the angle between the tangent of the disk and the horizontal: $\tan\theta = -f(X_j)/r$. We have $\theta = O(1/r)$. The horizontal velocity between rebounds is given by: $U_{j+1/2} = W_J \cos\theta - R_j \sin\theta$. From our previous estimates, we get: $W_j = O(\sqrt{\tfrac{g}{r}})$, $R_j = \sqrt{2E}[1 + O(r^{-2})]$, Then: $U_{j+1/2} = \left[W_j + \frac{\sqrt{2E}}{r}f(X_j)\right][1 + O(r^{-2})]$. Calling V'_j the vertical velocity after collision j, V_{j+1} the vertical velocity before collision $j+1$, and T the time between 2 rebounds, we have:

$$V'_j = \sqrt{2E}[1 + O(r^{-2})], \qquad V_{j+1} = -\sqrt{2E}[1 + O(r^{-2})],$$

$$T = \frac{2\sqrt{2E}}{g}[1 + O(r^{-2})].$$

Thus the mapping is given by:

$$X_{j+1} = X_j + \frac{2\sqrt{2E}}{g}U_{j+1/2}[1 + O(r^{-2})],$$

$$W_{j+1} = \left[U_{j+1/2} + \frac{\sqrt{2E}}{r}f(X_{j+1})\right][1 + O(r^{-2})].$$

Taking the limit $r \to \infty$ we finally obtain the mapping from (X_j, W_j) to (X_{j+1}, W_{j+1}):

$$U_{j+1/2} = W_j + \frac{\sqrt{2E}}{r}f(X_j),$$

$$X_{j+1} = X_j + \frac{2\sqrt{2E}}{g}U_{j+1/2},$$

$$W_{j+1} = U_{j+1/2} + \frac{\sqrt{2E}}{r}f(X_{j+1}).$$

Considering the vicinity of the fixed points $X = \pm 1$, $W = 0$, we introduce:

$$\cosh\phi = 1 + \frac{4E}{gr}, \qquad \sinh\phi = \sqrt{\frac{4E}{gr}\left(2 + \frac{4E}{gr}\right)},$$

$$U = u\sqrt{\frac{g}{2r}\left(2 + \frac{4E}{gr}\right)}, \quad W = w\sqrt{\frac{g}{2r}\left(2 + \frac{4E}{gr}\right)}.$$

The mapping in dimensionless form is:

$$u_{j+1/2} = w_j + f(X_j)\tanh\frac{\phi}{2},$$

$$X_{j+1} = X_j + u_{j+1/2}\sinh\phi,$$

$$w_{j+1} = u_{j+1/2} + f(X_{j+1})\tanh\frac{\phi}{2}.$$

ϕ is the only parameter left. It cannot be eliminated and is related to the eigenvalues of the fixed points. We therefore have achieved our goal of having tunable eigenvalues.

The mapping can be written in several forms, which can be useful in the following. We introduce: $x_j = f(X_j)$, $s_j = $ sign X_j and $x_j = X_j - s_j$. Eliminating $U_{j+1/2}$ from the equations, we get the mapping F:

$$X_{j+1} = X_j\cosh\phi + w_j\sinh\phi - s_j(\cosh\phi - 1),$$

$$w_{j+1} = X_j\sinh\phi + w_j\cosh\phi - (s_j\cosh\phi + s_{j+1})\tanh\frac{\phi}{2}.$$

Equivalently, we can write:

$$x_{j+1} = x_j\cosh\phi + w_j\sinh\phi + (s_j - s_{j+1}),$$

$$w_{j+1} = x_j\sinh\phi + w_j\cosh\phi - (s_j - s_{j+1})\tanh\frac{\phi}{2}.$$

or

$$x_{j+1} + w_{j+1} = e^{\phi}(x_j + w_j) + \frac{2e^{\phi}}{e^{\phi}+1}(s_j - s_{j+1}),$$

$$w_{j+1} - w_{j+1} = e^{-\phi}(x_j - w_j) + \frac{2e^{-\phi}}{e^{-\phi}+1}(s_j - s_{j+1}).$$

4.2   Properties of the motion

The inverse mapping F^{-1}:

$$x_j = x_{j+1}\cosh\phi - w_{j+1}\sinh\phi + (s_{j+1} - s_j),$$

$$w_j = -x_{j+1}\sinh\phi + w_{j+1}\cosh\phi - (s_{j+1} - s_j)\tanh\frac{\phi}{2}.$$

can be obtained from F by changing ϕ to $-\phi$. This results from the symmetry with respect to time of the original problem, and from the choice of x and W which preserves the symmetry.

Due to the very simple form of F, we obtain the explicit equations for the iterated mapping F^n:

$$
\begin{aligned}
x_n + w_n = e^{n\phi} \Bigg\{ & (x_0 + w_0) \\
& + \frac{2}{e^\phi + 1} \left[s_0 - (e^\phi - 1) \sum_{j=1}^{n-1} e^{-j\phi} s_j - e^{(1-n)\phi} s_n \right] \Bigg\},
\end{aligned}
$$

$$
\begin{aligned}
x_n - w_n = e^{-n\phi} \Bigg\{ & (x_0 - w_0) \\
& + \frac{2}{e^{-\phi} + 1} \left[s_0 + (1 - e^{-\phi}) \sum_{j=1}^{n-1} e^{j\phi} s_j - e^{(n-1)\phi} s_n \right] \Bigg\}.
\end{aligned}
$$

A note of caution is in order here about the meaning of the $\sum$ notation. The notation:

$$
\sum_{j=1}^{n} f(j)
$$

can be made valid not only for $n > 0$, but also for $n = 0$ and $n < 0$. The generalization of $\sum$, is done by analogy with integrals:

$$
\sum_{j=a+1}^{j=b} f(j) = \sum_{j=c}^{j=b} f(j) - \sum_{j=c}^{j=a} f(j),
$$

where c is an origin satisfying $c \le \min(a, b)$. This definition gives:

$$
\sum_{j=a+1}^{j=b} f(j) = \begin{cases} f(a+1) + f(a+2) + \cdots + f(b) & \text{if } b > a \\ 0 & \text{if } b = a \\ -f(b+1) - f(b+2) - \cdots - f(a) & \text{if } b < a \end{cases}
$$

We search the fixed points and note that in this case the rebounds take all place on the same disk. Hence:

$$
x_{j+1} = x_j \cosh \phi + w_j \sinh \phi, \qquad w_{j+1} = x_j \sinh \phi + w_j \cosh \phi.
$$

The fixed point is therefore $x = w = 0$. In the (X, w) plane, this translates in:

$$
\begin{cases} X = 1, & w = 0, & s = 1; \\ X = -1, & w = 0, & s = -1. \end{cases}
$$

From the previous equations for F, we have:

$$
x_{j+1} + w_{j+1} = (x_j + w_j)e^\phi, \qquad x_{j+1} - w_{j+1} = (x_j - w_j)e^{-\phi},
$$

from which we get the eigenvalues of the fixed points:

$$
\lambda = e^\phi, \qquad \lambda^{-1} = e^{-\phi},
$$

and the associated eigendirections:

$$w = x, \qquad w = -x.$$

Thus, the invariant manifolds are straight lines, interrupted at $X = 0$ (Fig. 20).

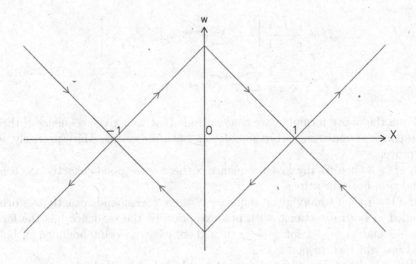

Fig. 20. Fixed points and invariant manifolds.

For given s_j and s_{j+1}, the mapping is always linear, with a homogeneous part given above. It follows that all orbits have the same Liapunov exponent, ϕ, and that all periodic orbits are unstable.

It is easy to show that there are five kinds of asymptotic regimes:

1. right-escaping orbit: $\quad X_j \to +\infty, \qquad w_j \to +\infty.$
2. right-asymptotic orbit: $\quad X_j \to +1, \qquad w_j \to 0.$
3. left-escaping orbit: $\quad X_j \to -\infty, \qquad w_j \to -\infty.$
4. left-asymptotic orbit: $\quad X_j \to -1, \qquad w_j \to 0.$
5. oscillating orbit: $\qquad X_j$ and w_j are bounded but have no limit.

To a given orbit, we associate the doubly infinite sequence:

$$S: \qquad \cdots s_{-2}, s_{-1}, s_0, s_1, s_2, \cdots$$

corresponding to the disks on which the particle bounces.

If an orbit is bounded for $j \to \infty$, then:

$$X_j + w_j = \frac{e^\phi - 1}{e^\phi + 1}\left[s_j + 2 \sum_{k=j+1}^{+\infty} e^{(j-k)\phi} s_k \right].$$

If an orbit is bounded for $j \to -\infty$, then:

$$X_j - w_j = \frac{1 - e^{-\phi}}{e^{-\phi} + 1} \left[s_j + 2 \sum_{k=-\infty}^{j-1} e^{(k-j)\phi} s_k \right].$$

If an orbit is bounded in both directions, then:

$$X_j = \frac{e^{\phi} - 1}{e^{\phi} + 1} \left[s_j + \sum_{k=1}^{+\infty} e^{-k\phi}(s_{j+k} + s_{j-k}) \right],$$

$$w_j = \frac{e^{\phi} - 1}{e^{\phi} + 1} \sum_{k=1}^{+\infty} e^{-k\phi}(s_{j+k} - s_{j-k}).$$

From the above formulae, we can conclude that to a given sequence S there corresponds at most one orbit bounded in both directions. More precisely, we can prove:

If $e^{\phi} > 3$ then to any given sequence S there corresponds exactly one orbit bounded in both directions.

If $e^{\phi} = 3$ then to any given sequence S there corresponds exactly one orbit bounded in both directions, with one exception: if the sequence has the form $s_j = -1$ and $s_k = +1$ for $k \neq j$, then there exist no orbit bounded in both directions which corresponds to it.

If $e^{\phi} < 3$ then there exist sequences S to which no bounded orbit corresponds.

In the particular case of the one-parameter family we have chosen the inital conditions are:

$$X_0 = h, \qquad u_{1/2} = 0,$$

with $-\infty < h < +\infty$. These orbits are called the h-orbits. This gives:

$$w_o = -(h - s_0) \tanh \frac{\phi}{2}, \qquad X_1 = h, \qquad w_1 = (h - s_0) \tanh \frac{\phi}{2}.$$

The orbit is symmetrical with respect to the $w = 0$ axis, so:

$$X_j = X_{1-j}, \qquad s_j = s_{1-j}, \qquad w_j = -w_{1-j}.$$

The iterated mapping F^n becomes:

$$x_n + w_n = \frac{2e^{n\phi}}{e^{\phi} + 1} \left[h - (e^{\phi} - 1) \sum_{j=1}^{n-1} e^{-j\phi} s_j - e^{(1-n)\phi} s_n \right],$$

$$x_n - w_n = \frac{2e^{-n\phi}}{e^{-\phi} + 1} \left[h + (1 - e^{-\phi}) \sum_{j=1}^{n-1} e^{j\phi} s_j - e^{(n-1)\phi} s_n \right].$$

4.3   Symbolic dynamics

To a given h-orbit we can associate a sequence of binary digits

$$D: \qquad d_1, d_2, \cdots$$

with $d_j = 0$ if $s_j = -1$ and $d_j = 1$ if $s_j = 1$, for $j = 1, 2, \cdots$) The D sequence differs from the S sequence in that the d_j are defined only for $j > 0$.

Then we define a number A by its binary representation:

$$A = 0.d_1 d_2 d_3 \cdots = \sum_{j=1}^{\infty} d^{-j} d_j.$$

Clearly, $0 \leq A \leq 1$. A given sequence defines one value of A, but there might be *two* sequences with the same value of A.

$A = k \times 2^{-n}$ with k and n integers; this is called a round number.

1. If $0 < A < 1$, then there exists one representation of A for which k is odd and $n > 0$. There are 2 different D sequences corresponding to A: $0.d_1 d_2 \cdots d_{n-1} 0111 \cdots$ and $0.d_1 d_2 \cdots d_{n-1} 1000 \cdots$.
2. If $A = 0$, there is only 1 corresponding D sequence: $0.000 \cdots$
3. If $A = 1$, there is only 1 corresponding D sequence: $0.111 \cdots$.

These sequences are either 0-ending or 1-ending.

A is not of the previous form: this is called a non-round number. There is always exactly one corresponding D sequence. It is neither 0-ending nor 1-ending. This is called an oscillating sequence.

Because of the symmetry of the h-orbits, the asymptotic behaviour of X_j is the same for $j \to +\infty$ and for $j \to -\infty$, which justifies that we use the D sequences instead of the S sequences.

There is a simple correspondence between the types of orbits, the D sequence and A.

orbit	D sequence	A
right-escaping	1-ending	round
right-asymptotic	1-ending	round
left-escaping	0-ending	round
left-asymptotic	0-ending	round
oscillating	oscillating	non-round

In a continuity interval, the orbit changes continuously, so A is constant. This suggests to look at the function $A(h)$. Fig. 21 shows the numerical result for $\lambda = e^{\Phi} = 3.5$. The reader will have recognized a *Devil's staircase* with an infinite number of horizontal steps. In the case where $e^{\Phi} \geq 3$, A is a continuous, non-decreasing function of h, and it is possible to explain completely the structure.

In the case where $e^{\Phi} < 3$ (Fig. 22), the structure is more complex and we cannot fully analyse it. We can still show that A is a non-decreasing function of h.

At this point, we study the inverse problem: which values of h correspond to a given D sequence, or to a given A? We have to consider different type of values of A.

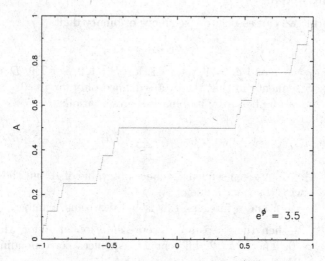

Fig. 21. The function $A(h)$ for $e^\phi = 3.5$.

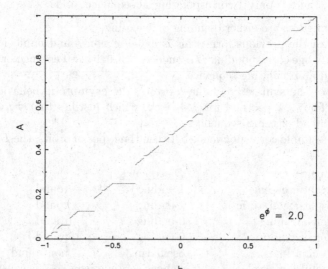

Fig. 22. The function $A(h)$ for $e^\phi = 2.0$.

- Non-round A: there corresponds at most 1 h-orbit.
  1. If $e^\phi \geq 3$, there corresponds exactly 1 h-orbit.

$$h = (e^\phi - 1) \sum_{j=1}^{+\infty} e^{-j\phi} s_j.$$

2. If $e^\phi < 3$, there exist non-round A to which no h-orbit corresponds.

- Round A, $0 < A < 1$: there are 2 corresponding D sequences.
  1. Asymptotic orbits. To a given 0-ending or 1-ending sequence there corresponds at most 1 asymptotic h-orbit. If $e^\phi \geq 3$, there corresponds exactly 1 h-orbit. h is given by the previous formula. Since D is of the form s_1 to s_n arbitrary; $s_j = -s_n$ for $j > n$, we can write:

$$h = (e^\phi - 1) \sum_{j=1}^{n-1} e^{-j\phi} s_j + (e^\phi - 2)e^{-n\phi} s_n.$$

  2. Escaping orbits. The 1-ending sequence corresponds to a right-asymptotic orbit:

$$h_- = (e^\phi - 1) \sum_{j=1}^{n-1} e^{-j\phi} s_j - (e^\phi - 2)e^{-n\phi}.$$

  The 0-ending sequence corresponds to a left–asymptotic orbit:

$$h_+ = (e^\phi - 1) \sum_{j=1}^{n-1} e^{-j\phi} s_j + (e^\phi - 2)e^{-n\phi}.$$

  These two values satisfy the relation:

$$h_+ - h_- = 2(e^\phi - 2)e^{-n\phi} > 0.$$

  Since $A(h)$ is non-decreasing, the whole interval $h_- \leq h \leq h_+$ corresponds to the same value of A. The open interval $h_- < h < h_+$ consists entirely of escaping orbits. The interval of h values corresponding to this value of A does not extend past h_- or h_+.
- $A = 0$: there is only h_+, but no h_-. This value corresponds to the interval $-\infty < h \leq -1$. $h = -1$ corresponds to a left-asymptotic orbit: the left fixed point of F. The open interval $-\infty < h < -1$ corresponds to a left-escaping orbit.
- $A = 1$: there is only h_-, but no h_+. This value corresponds to the interval $+1 \leq h < +\infty$. $h = +1$ corresponds to a right-asymptotic orbit: the right fixed point of F. The open interval $+1 < h < +\infty$ corresponds to a right-escaping orbit.

Provided $e^\phi \geq 3$, there are two additional results that can be proved:

- The curve $A(h)$ has exact self-similarity. The curve as a whole extends from $h = -1$ to $h = +1$ and from $A = 0$ to $A = 1$. In the lower left corner is an exact replica of the whole picture, reduced by a factor e^Φ horizontally and 2 vertically, extending from $h = -1$ to $h = -1 + 2e^{-\Phi}$ and from $A = 0$ to $A = 1/2$. There is an identical replica in the upper right corner.

- The set of values of h corresponding to bounded orbits forms a Cantor set, with measure 0 and with fractal dimension

$$\ln(2)/\Phi.$$

In the borderline case $e^{\Phi} = 3$, we obtain exactly the classical Cantor set (repeated exclusion of the middle third). The asymptotic orbits form an enumerable subset of the bounded orbits; this subset also has the dimension given above.

References

1. N. Agmon: J. Chem. Phys. **76**, 1309 (1982)
2. R. Blümel, U. Smilansky: Phys. Rev. Lett. **60**, 477 (1988)
3. S.F. Dermott, C.D. Murray: Icarus **48**, 1 (1981)
4. B.Eckhardt, H. Aref: Phil. Trans. R. Soc. Lond. A **326**, 655 (1989)
5. B. Eckhardt, C. Jung: J. Phys. A **19**, L829 (1986)
6. D.E. Fitz, P. Brumer: J. Chem. Phys. **70**, 5527 (1979)
7. P. Goldreich, S. Tremaine: Nature **277**, 97 (1979)
8. P. Goldreich, S. Tremaine: Astrophys. J. **241**, 425 (1980)
9. L. Gottdiener: Molecular Physics **29**, 1585 (1975)
10. M. Hénon: Astron. Astrophys. **1**, 223 (1969)
11. M. Hénon: Physica D **33**, 132 (1988)
12. M. Hénon, J.-M. Petit: Celestial Mechanics **38**, 67 (1986)
13. C. Jung, H.-J. Scholz: J. Phys. A **20**, 3607 (1987)
14. S.V. Manakov, L.N. Shchur: JETP Lett. **37**, 54 (1983)
15. C. Marchal: J. Differ. Equations **23**, 387 (1977)
16. J. Moser: *Stable and Random Motions in Dynamical Systems with Special Emphasys on Celestial Mechanics* (Princeton University Press, 1973)
17. D.W. Noid, S.K. Gray, S.A. Rice: J. Chem. Phys. **84**, 2649 (1986)
18. J.-M. Petit, M. Hénon: Icarus **66**, 536 (1986)
19. C.C. Rankin, W.H. Miller: J. Chem. Phys. **55**, 3150 (1971)
20. C.G. Schlier: Chemical Physics **77**, 267 (1983)
21. C.F. Yoder, G. Colombo, S.P. Synnott, K.A. Yoder: Icarus **53**, 431 (1983)

Close Encounters in Öpik's Theory

Giovanni B. Valsecchi

Istituto di Astrofisica Spaziale
Area di Ricerca del C.N.R.
via Fosso del Cavaliere 100 00133 Roma (Italy)
e-mail: giovanni@ias.rm.cnr.it

Abstract. In many cases of interest, close encounters of small bodies and planets can be treated analytically in the framework of a theory due to Öpik, that has been used in many statistical studies of close encounters. The basic formulation of the theory can be extended by the explicit introduction of the nodal distance and of time; it is then possible to reproduce the basic features of the encounters with the Earth of some near-Earth asteroids, and to uncover some geometric properties of resonant orbits. The variables used in Öpik's theory turn out to be useful also in the problem of the identification of meteoroid streams.

1   Introduction

Close encounters with the planets are known to be the cause of fast orbital evolution for those small solar system bodies that are either in planet-crossing orbits or in non-planet-crossing orbits that anyway allow strong gravitational interactions with a planet.

In the latter case the motion in the vicinity of the planet is rather complicated, and temporary satellite captures are possible. Many cases of this type of behaviour have been found in studies of the motion of Jupiter-family comets; the best known cases are those of 39P/Oterma [4] [5] [6], of 82P/Gehrels 3 [21] [3] [6], of 111P/Helin-Roman-Crockett [8] [23], and of D/Shoemaker-Levy 9 [1] [15]; actually, numerical studies of motion of D/Shoemaker-Levy 9 have shown that the comet presumably underwent a very long satellite capture before the collision with Jupiter.

On the other hand, when the planetocentric velocity of the small body is sufficiently high, so that the planetocentric orbit is hyperbolic and the encounter duration is short, the quantitative description of the motion is greatly simplified [9], and can be done using Öpik's theory of close encounters [20].

This theory allows, under rather simple assumptions, to treat close encounters of small bodies and planets analytically. In its standard formulation, the motion of the bodies is taken as rectilinear near the encounter, and nothing is said about the distance between the two trajectories, implicitly assumed to be small, but never appearing into the equations. In this formulation the theory has been used only to get qualitative results, or to do statistical investigations.

If the minimum distance between the trajectories is explicitly introduced [26], together with a time coordinate, it is possible to have explicit expressions for the motion of the small body, thus making a comparison with a numerical integration of the actual motion feasible. Moreover, all the post-encounter parameters become analytically computable from the initial conditions.

We will draw particular attention to the perturbations of the semimajor axis of the orbit of the small body; the distribution of energy perturbations of a given ensemble of initial orbits can be computed analytically [25], and it is shown that all the initial conditions leading to the same perturbation of the semimajor axis are characterized by a very simple geometric property.

Furthermore, the variables used in Öpik's theory turn out to be ideally suited to describe meteor orbits [24], as they are in one-to-one correspondence with the quantities actually observed, and can in fact be used to search for meteoroid streams [14]; moreover, two of these quantities are invariant for the principal secular perturbation affecting meteoroid streams.

2   Basic formulae of Öpik's theory

2.1   The components of the planetocentric velocity

The angle α between the position and velocity vectors can be computed from the expression of the angular momentum L:

$$r \quad = \frac{a(1 - e^2)}{1 + e \cos f} \tag{1}$$

$$v \quad = \sqrt{\frac{2}{r} - \frac{1}{a}} = \sqrt{\frac{2a - r}{ra}} \tag{2}$$

$$L \quad = \sqrt{a(1 - e^2)} = rv \sin \alpha = r\sqrt{\frac{2a - r}{ra}} \sin \alpha \tag{3}$$

$$\sin \alpha = \sqrt{\frac{a^2(1 - e^2)}{2ar - r^2}}. \tag{4}$$

Setting $r = 1$:

$$\sin \alpha = \sqrt{\frac{a^2(1 - e^2)}{2a - 1}}. \tag{5}$$

Now, set the reference frame so that the small body is in $(1, 0, 0)$ (this implies it to be at one of the nodes of its orbit); its heliocentric velocity will have components:

$$\begin{pmatrix} v_x \\ v_y \\ v_z \end{pmatrix} = \begin{pmatrix} v \cos \alpha \\ v \sin \alpha \cos i \\ \pm v \sin \alpha \sin i \end{pmatrix} = \begin{pmatrix} \pm\sqrt{2 - 1/a - a(1 - e^2)} \\ \sqrt{a(1 - e^2)} \cos i \\ \pm\sqrt{a(1 - e^2)} \sin i \end{pmatrix}. \tag{6}$$

To obtain the planetocentric velocity U, for a planet on a circular orbit at unit distance from the Sun, and for which we disregard the effect of the mass on the velocity, we simply subtract its velocity, that has components $(0, 1, 0)$:

$$\begin{pmatrix} U_x \\ U_y \\ U_z \end{pmatrix} = \begin{pmatrix} \pm\sqrt{2 - 1/a - a(1 - e^2)} \\ \sqrt{a(1 - e^2)} \cos i - 1 \\ \pm\sqrt{a(1 - e^2)} \sin i \end{pmatrix}; \tag{7}$$

its modulus is:

$$U = \sqrt{3 - \frac{1}{a} - 2\sqrt{a(1 - e^2)}\cos i}. \tag{8}$$

This can be rewritten as:

$$U = \sqrt{3 - T} \tag{9}$$

where T is the Tisserand parameter with respect to the planet:

$$T = \frac{1}{a} + 2\sqrt{a(1 - e^2)}\cos i. \tag{10}$$

2.2 The angles θ and ϕ

We can introduce two angles, θ and ϕ, such that (Fig. 1):

$$\begin{pmatrix} U_x \\ U_y \\ U_z \end{pmatrix} = \begin{pmatrix} U\sin\theta\sin\phi \\ U\cos\theta \\ U\sin\theta\cos\phi \end{pmatrix} \tag{11}$$

and, conversely:

$$\cos\theta = \frac{U_y}{U} \tag{12}$$

$$\tan\phi = \frac{U_x}{U_z} \tag{13}$$

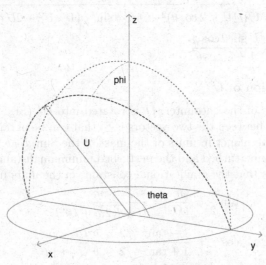

Fig. 1. The geometry of the pre-encounter planetocentric velocity vector U.

Using the previous expressions, θ, ϕ can be obtained in terms of a, e, i:

$$\cos \theta = \frac{\sqrt{a(1 - e^2)} \cos i - 1}{\sqrt{3 - 1/a - 2\sqrt{a(1 - e^2)} \cos i}} \tag{14}$$

$$\tan \phi = \frac{\pm \sqrt{2 - 1a - a(1 - e^2)}}{\pm \sqrt{a(1 - e^2)} \sin i}. \tag{15}$$

Since

$$\sqrt{a(1 - e^2)} \cos i = \frac{3 - 1/a - U^2}{2}, \tag{16}$$

a simple expression for θ is in terms of U, a:

$$\cos \theta = \frac{1 - U^2 - 1/a}{2U}. \tag{17}$$

Finally, a, e, i can be obtained from U_x, U_y, U_z:

$$a = \frac{1}{1 - U^2 - 2U_y} \tag{18}$$

$$e = \sqrt{U^4 + 4U_y^2 + U_x^2(1 - U^2 - 2U_y) + 4U^2 U_y} \tag{19}$$

$$\tan i = \frac{U_z}{1 + U_y} \tag{20}$$

and from U, θ, ϕ:

$$a = \frac{1}{1 - U^2 - 2U \cos \theta} \tag{21}$$

$$e = U\sqrt{(U + 2\cos \theta)^2 + \sin^2 \theta \sin^2 \phi (1 - U^2 - 2U \cos \theta)} \tag{22}$$

$$\tan i = \frac{U \sin \theta \cos \phi}{1 + U \cos \theta} \tag{23}$$

2.3   The rotation of U

As a consequence of the encounter, U is rotated into U' (Fig. 2), of the same length; the angle between the two vectors is γ, that can be obtained in terms of m, the mass of the planet in units of the mass of the Sun, U, b, and d, the two latter being the unperturbed and the perturbed minimum distance at encounter, respectively (note that the gravitational constant, in the units used, is 1):

$$\tan \frac{\gamma}{2} = \frac{m}{bU^2} = \frac{m}{U\sqrt{d(2m + U^2 d)}} \tag{24}$$

$$\cos \gamma = \frac{1 - \tan^2 \gamma/2}{1 + \tan^2 \gamma/2} = \frac{b^2 U^4 - m^2}{b^2 U^4 + m^2} \tag{25}$$

$$\sin \gamma = \frac{2 \tan \gamma/2}{1 + \tan^2 \gamma/2} = \frac{2mbU^2}{b^2 U^4 + m^2} \tag{26}$$

$$\sin \frac{\gamma}{2} = \frac{m}{\sqrt{m^2 + b^2 U^4}} = \frac{m}{m + U^2 d}. \tag{27}$$

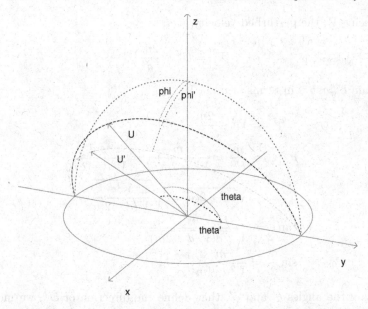

Fig. 2. The geometry of the rotation of the pre-encounter planetocentric velocity vector U into the post-encounter vector U'. Some angles are not evidenced, in order not to clutter the plot. The angle between U and U' is γ, that is also one of the sides of the spherical triangle whose remaining sides are θ and θ' (see text). In this triangle, $\chi = \phi - \phi'$ is the angle opposite to γ, and ψ is the angle opposite to θ'.

These expressions can be simplified putting $c = \dfrac{m}{U^2}$:

$$\tan\frac{\gamma}{2} = \frac{c}{b} = \frac{Uc}{\sqrt{d(2m + U^2 d)}} \tag{28}$$

$$\cos\gamma = \frac{b^2 U^4 - m^2}{b^2 U^4 + m^2} = \frac{b^2 - c^2}{b^2 + c^2} \tag{29}$$

$$\sin\gamma = \frac{2mbU^2}{b^2 U^4 + m^2} = \frac{2bc}{b^2 + c^2} \tag{30}$$

$$\sin\frac{\gamma}{2} = \frac{m}{m + U^2 d} = \frac{U^2 c}{U^2 c + U^2 d} = \frac{c}{c + d}. \tag{31}$$

Furthermore, b can be expressed in terms of m, U, d:

$$b = \frac{\sqrt{d(2m + U^2 d)}}{U} = \sqrt{d^2 + 2cd}, \tag{32}$$

and d in terms of m, U, b:

$$d = \sqrt{\frac{m^2}{U^4} + b^2} - \frac{m}{U^2} = \sqrt{b^2 + c^2} - c; \tag{33}$$

introducing V, the perturbed velocity at d:

$$V^2 = U^2 + \frac{2m}{d} \tag{34}$$

we obtain U, c, b γ in terms of m, d, V:

$$U^2 \quad = V^2 - \frac{2m}{d} \tag{35}$$

$$c \quad = \frac{m}{U^2} = \frac{m}{V^2 - 2m/d} = \frac{md}{V^2 d - 2m} \tag{36}$$

$$b \quad = \sqrt{d^2 + 2cd} = d\sqrt{\frac{V^2 d}{V^2 d - 2m}} \tag{37}$$

$$\tan\frac{\gamma}{2} = \frac{m}{V\sqrt{d(V^2 d - 2m)}} \tag{38}$$

$$\sin\frac{\gamma}{2} = \frac{m}{V^2 d - m}. \tag{39}$$

To obtain the angles θ' and ϕ', that define the direction of $\boldsymbol{U'}$, we must first introduce the components of $\boldsymbol{b}$, i.e. the vector going from the planet to the coordinates of the intersection of the direction of $\boldsymbol{U}$ with the plane perpendicular to $\boldsymbol{U}$ and containing the planet; this b-plane is the ξ-ζ plane of a ξ-η-ζ reference frame whose η-axis is directed along $\boldsymbol{U}$, and ξ and ζ are oriented as specified in the next section. We define, in agreement with [7],

$$\xi = b\sin\psi \tag{40}$$
$$\zeta = b\cos\psi; \tag{41}$$

then θ' and ϕ' can be obtained in terms of θ, ϕ, γ, ψ. In particular, let us consider the spherical triangle with sides γ, θ and θ', in which $\chi = \phi - \phi'$ is the angle opposite to γ and ψ is the angle opposite to θ'; for the law of cosines for sides:

$$\cos\theta' = \cos\theta\cos\gamma + \sin\theta\sin\gamma\cos\psi, \tag{42}$$

and for the law of sines:

$$\sin\chi = \frac{\sin\psi\sin\gamma}{\sin\theta'}. \tag{43}$$

Moreover, the law of cosines for sides applied again gives:

$$\cos\chi = \frac{\cos\gamma - \cos\theta\cos\theta'}{\sin\theta\sin\theta'}$$
$$= \frac{\sin\theta\cos\gamma - \cos\theta\sin\gamma\cos\psi}{\sin\theta'}, \tag{44}$$

so that:

$$\tan\chi = \frac{\sin\gamma\sin\psi}{\sin\theta\cos\gamma - \cos\theta\sin\gamma\cos\psi} \tag{45}$$

$$\tan \phi' = \frac{\tan \phi - \tan \chi}{1 + \tan \phi \tan \chi}$$
$$= \frac{\sin \theta \tan \phi \cos \gamma - \cos \theta \tan \phi \sin \gamma \cos \psi - \sin \gamma \sin \psi}{\sin \theta \cos \gamma - \cos \theta \sin \gamma \cos \psi + \tan \phi \sin \gamma \sin \psi} \tag{46}$$

$$\cos \phi' = \frac{1}{\sqrt{1 + \tan^2 \phi'}}$$
$$= \frac{\cos \phi (\sin \theta \cos \gamma - \cos \theta \sin \gamma \cos \psi) + \sin \phi \sin \gamma \sin \psi}{\sqrt{(\sin \theta \cos \gamma - \cos \theta \sin \gamma \cos \psi)^2 + \sin^2 \gamma \sin^2 \psi}} \tag{47}$$

$$\sin \phi' = \cos \phi' \tan \phi'$$
$$= \frac{\sin \phi (\sin \theta \cos \gamma - \cos \theta \sin \gamma \cos \psi) - \cos \phi \sin \gamma \sin \psi}{\sqrt{(\sin \theta \cos \gamma - \cos \theta \sin \gamma \cos \psi)^2 + \sin^2 \gamma \sin^2 \psi}}. \tag{48}$$

3   From the planetocentric to the b-plane frame and back

The transformation from the planetocentric frame (axes X, Y, Z, with the Sun on the negative X-axis and the Y-axis in the direction of the motion of the planet) to that of the b-plane (axes ξ, η, ζ) is accomplished by the following two rotations:

- by $-\phi$ (i.e. by ϕ in the clockwise direction) about Y;
- by $-\theta$ about ξ (which is perpendicular to the old Y-axis and to U);

in column notation:

$$\begin{pmatrix} \xi \\ \eta \\ \zeta \end{pmatrix} = \begin{pmatrix} 1 & 0 & 0 \\ 0 & \cos \theta & \sin \theta \\ 0 & -\sin \theta & \cos \theta \end{pmatrix} \begin{pmatrix} \cos \phi & 0 & -\sin \phi \\ 0 & 1 & 0 \\ \sin \phi & 0 & \cos \phi \end{pmatrix} \begin{pmatrix} X \\ Y \\ Z \end{pmatrix}$$
$$= \begin{pmatrix} X \cos \phi - Z \sin \phi \\ (X \sin \phi + Z \cos \phi) \sin \theta + Y \cos \theta \\ (X \sin \phi + Z \cos \phi) \cos \theta - Y \sin \theta \end{pmatrix}. \tag{49}$$

The inverse transformation is accomplished by the following two rotations:

- by θ about ξ;
- by ϕ about Y;

in column notation:

$$\begin{pmatrix} X \\ Y \\ Z \end{pmatrix} = \begin{pmatrix} \cos \phi & 0 & \sin \phi \\ 0 & 1 & 0 \\ -\sin \phi & 0 & \cos \phi \end{pmatrix} \begin{pmatrix} 1 & 0 & 0 \\ 0 & \cos \theta & -\sin \theta \\ 0 & \sin \theta & \cos \theta \end{pmatrix} \begin{pmatrix} \xi \\ \eta \\ \zeta \end{pmatrix}$$
$$= \begin{pmatrix} (\eta \sin \theta + \zeta \cos \theta) \sin \phi + \xi \cos \phi \\ \eta \cos \theta - \zeta \sin \theta \\ (\eta \sin \theta + \zeta \cos \theta) \cos \phi - \xi \sin \phi \end{pmatrix}. \tag{50}$$

To get a better idea of the geometry, let us consider the intersection of the b-plane with the ecliptic, and the projections on the b-plane of X, Y, and Z.

3.1   The ecliptic on the b-plane

On the ecliptic $Z = 0$, so that:

$$\begin{pmatrix} \xi \\ \eta \\ \zeta \end{pmatrix} = \begin{pmatrix} X \cos\phi \\ X \sin\phi \sin\theta + Y \cos\theta \\ X \sin\phi \cos\theta - Y \sin\theta \end{pmatrix} ; \tag{51}$$

on the b-plane $\eta = 0$, and therefore:

$$0 = X \sin\phi \sin\theta + Y \cos\theta \rightarrow Y = -X \frac{\sin\phi \sin\theta}{\cos\theta}$$

$$\begin{pmatrix} \xi \\ \zeta \end{pmatrix} = \begin{pmatrix} X \cos\phi \\ X \sin\phi \cos\theta - Y \sin\theta \end{pmatrix} = \begin{pmatrix} X \cos\phi \\ X(\sin\phi/\cos\theta) \end{pmatrix} . \tag{52}$$

3.2   The projection of the X-axis on the b-plane

For $Y = Z = 0$ one has:

$$\begin{pmatrix} \xi \\ \eta \\ \zeta \end{pmatrix} = \begin{pmatrix} X \cos\phi \\ X \sin\phi \sin\theta \\ X \sin\phi \cos\theta \end{pmatrix} ; \tag{53}$$

and therefore:

$$\begin{pmatrix} \xi \\ \zeta \end{pmatrix} = \begin{pmatrix} X \cos\phi \\ X \sin\phi \cos\theta \end{pmatrix} . \tag{54}$$

3.3   The projection of the Y-axis on the b-plane

For $X = Z = 0$ one has:

$$\begin{pmatrix} \xi \\ \eta \\ \zeta \end{pmatrix} = \begin{pmatrix} 0 \\ Y \cos\theta \\ -Y \sin\theta \end{pmatrix} ; \tag{55}$$

and therefore:

$$\begin{pmatrix} \xi \\ \zeta \end{pmatrix} = \begin{pmatrix} 0 \\ -Y \sin\theta \end{pmatrix} . \tag{56}$$

Note that the ζ-axis is always the projection of the Y-axis (actually, of the $-Y$-axis). Since a difference in the timing of encounter between the small body and the planet would mean that the position of the latter, at the time when the small body crosses the ecliptic, would change along the Y-axis, this implies that the ζ-axis can be seen as a sort of time axis, and in particular that a distribution of fictitious small bodies all on the same orbit, separated only in mean anomaly, would show up in the b-plane as a segment parallel to the ζ-axis.

3.4   The projection of the Z-axis on the b-plane

For $X = Y = 0$ one has:

$$\begin{pmatrix} \xi \\ \eta \\ \zeta \end{pmatrix} = \begin{pmatrix} -Z \sin \phi \\ Z \cos \phi \sin \theta \\ Z \cos \phi \cos \theta \end{pmatrix} ; \tag{57}$$

and therefore:

$$\begin{pmatrix} \xi \\ \zeta \end{pmatrix} = \begin{pmatrix} -Z \sin \phi \\ Z \cos \phi \cos \theta \end{pmatrix} . \tag{58}$$

3.5   Examples

Figure 3 shows the intersection with the ecliptic and the projections of X, Y, and Z on the b-plane of the October 2028 encounter with the Earth of asteroid 1997XF$_{11}$; this encounter will take place at the descending node, in the post-perihelion branch of the orbit. The orbit of 1997XF$_{11}$ has $a = 1.442$ AU, $e = 0.484$ and $i = 4.1°$, $U = 0.459$, $\theta = 84.0°$, and $\phi = 99.5°$.

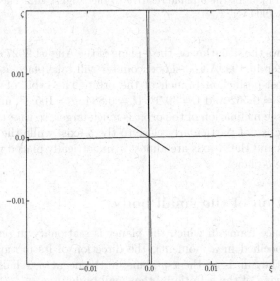

Fig. 3. The intersection of the ecliptic and the projections of X, Y, and Z on the b-plane of the October 2028 encounter with the Earth of asteroid 1997XF$_{11}$. The lengths of all the axes drawn is equal to the semi-width of the plot (0.02 AU), in order to give an idea of the spatial arrangement. The positive end of the X-axis is denoted by a large full dot, the positive end of the Y-axis has a small dot (note it on the negative ζ-axis), and the positive end of the Z-axis has an open circle. The uninterrupted line is the intersection with the ecliptic.

As it is possible to see, for this low-inclination asteroid the intersection with the ecliptic is very close to the ζ-axis, while the projection of the Z-axis is very close to the ξ-axis of the b-plane.

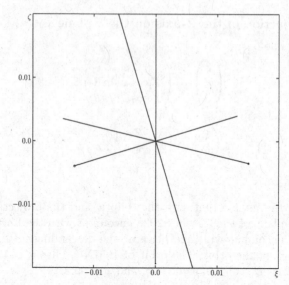

Fig. 4. Same as Fig. 3 for the b-plane relative to the August 2027 encounter with the Earth of asteroid 1999AN$_{10}$.

Figure 4 shows the situation on the b-plane of the August 2027 encounter with the Earth of asteroid 1999AN$_{10}$; this encounter will take place at the ascending node, in the post-perihelion branch of the orbit. The orbit of 1999AN$_{10}$ has $a = 1.459$ AU, $e = 0.562$ and $i = 39.9°$, $U = 0.884$, $\theta = 105.3°$, and $\phi = 41.3°$.

In this case the inclination of the orbit is much larger, so that the intersection with the ecliptic is not particularly close to the ζ-axis, while the projections of both the X-axis and the Z-axis are almost symmetrically placed with respect to the ξ-axis of the b-plane.

4   The motion of the small body

We use a reference frame in which the planet is stationary in one point of the y-axis (to be specified in a moment), the direction of its motion is along the same axis, and the Sun is on the x-y plane, in the negative-x half-plane. If t_0 is the time of crossing of the ecliptic by the small body:

$$\begin{pmatrix} x(t) \\ y(t) \\ z(t) \end{pmatrix} = \begin{pmatrix} U_x(t - t_0) + x(t_0) \\ U_y(t - t_0) + y(t_0) \\ U_z(t - t_0) \end{pmatrix}. \tag{59}$$

We set $y(t_0) = 0$, and $x(t_0) = x_0$ (the nodal distance), so that:

$$\begin{pmatrix} x(t) \\ y(t) \\ z(t) \end{pmatrix} = \begin{pmatrix} U \sin\theta \sin\phi(t - t_0) + x_0 \\ U \cos\theta(t - t_0) \\ U \sin\theta \cos\phi(t - t_0) \end{pmatrix}, \tag{60}$$

and the coordinates of the planet are $(0, y_p, 0)$.

4.1   The local Minimum Orbital Intersection Distance (MOID)

In many cases of interest, close planetary encounters are possible if one or both of the nodes of the orbit of the small body are close to the orbit of a planet. In this case, it is useful to speak of *local* Minimum Orbital Intersection Distance (MOID), indicating with this expression the local minimum of the distance between the two orbits.

In the framework of Öpik's theory it is possible to deduce an approximate expression for the local MOID at a specified node, by considering that the small body travels on a straight line:

$$\begin{pmatrix} x \\ z \end{pmatrix} = \begin{pmatrix} (U_x/U_y)y + x_0 \\ (U_z/U_y)y \end{pmatrix};\tag{61}$$

then, the square of the distance from the y-axis is:

$$D_y^2 = x^2 + z^2 = \frac{U_x^2 + U_z^2}{U_y^2}y^2 + 2\frac{U_x}{U_y}x_0 y + x_0^2\tag{62}$$

and its derivative is:

$$\frac{d(D_y^2)}{dy} = \frac{2(U_x^2 + U_z^2)}{U_y^2}y + \frac{2U_x}{U_y}x_0;\tag{63}$$

it is zero at:

$$y = -\frac{U_x U_y}{U_x^2 + U_z^2}x_0.\tag{64}$$

In terms of θ and ϕ we have:

$$y = -\frac{\cos\theta\sin\phi}{\sin\theta}x_0,\tag{65}$$

so that:

$$D_y^2 = x_0^2\cos^2\phi.\tag{66}$$

This means that the ratio between the local MOID, close to the node under examination, and x_0 is:

$$D_y/x_0 = \cos\phi.\tag{67}$$

This expression can be generalized to the case of an elliptic orbit for the planet [2].

In order to have an encounter at the minimum possible distance the planet (stationary in our reference frame) must be at $(0, -\cos\theta(\sin\phi/\sin\theta)x_0, 0)$, and the small body must pass at the MOID point at time t_{MOID}:

$$y_{MOID} = -\frac{U_x U_y}{U_x^2 + U_z^2}x_0 = U_y(t_{MOID} - t_0)\tag{68}$$

$$t_{MOID} - t_0 = -\frac{U_x}{U_x^2 + U_z^2}x_0 = -\frac{\sin\phi}{U\sin\theta}x_0.\tag{69}$$

At $t = t_{MOID}$ the coordinates of the small body are:

$$\begin{pmatrix} x(t_{MOID}) \\ y(t_{MOID}) \\ z(t_{MOID}) \end{pmatrix} = \begin{pmatrix} U \sin\theta \sin\phi(t_{MOID} - t_0) + x_0 \\ U \cos\theta(t_{MOID} - t_0) \\ U \sin\theta \cos\phi(t_{MOID} - t_0) \end{pmatrix}$$
$$= \begin{pmatrix} x_0 \cos^2\phi \\ -x_0 \cos\theta \sin\phi/\sin\theta \\ -x_0 \sin\phi \cos\phi \end{pmatrix} \tag{70}$$

and the distance is

$$D_y = \sqrt{x_0^2 \cos^4\phi + x_0^2 \sin^2\phi \cos^2\phi} = x_0 \cos\phi. \tag{71}$$

We can pass from (x, y, z) to the planetocentric axes (X, Y, Z):

$$\begin{pmatrix} X \\ Y \\ Z \end{pmatrix} = \begin{pmatrix} x \\ y + (\cos\theta \sin\phi/\sin\theta)x_0 \\ z \end{pmatrix} \tag{72}$$

so that the coordinates on the b-plane are:

$$\begin{pmatrix} \xi \\ \eta \\ \zeta \end{pmatrix} = \begin{pmatrix} X\cos\phi - Z\sin\phi \\ (X\sin\phi + Z\cos\phi)\sin\theta + Y\cos\theta \\ (X\sin\phi + Z\cos\phi)\cos\theta - Y\sin\theta \end{pmatrix} = \begin{pmatrix} x_0\cos\phi \\ 0 \\ 0 \end{pmatrix}. \tag{73}$$

This result is especially interesting in view of the fact, already remarked before, that displacements along the ζ-axis correspond to changes in the time of encounter; now we see that displacements along the ξ-axis correspond to changes in the distance between the orbits. Thus, the two key factors governing the possibility of a collision, i.e. orbital distance and time separation at encounter, map nicely in the two axes that we have defined on the b-plane.

4.2   The planetocentric orbital elements of the small body

The planetocentric semimajor axis can be obtained from the energy at infinity

$$a_g = -\frac{m}{U^2} = -c \tag{74}$$

and the eccentricity from the magnitude of the angular momentum

$$e_g = \sqrt{1 - \frac{b^2 U^2}{a_g}} = \sqrt{1 + \frac{b^2}{c^2}} = \frac{\sqrt{b^2 + c^2}}{c}. \tag{75}$$

To compute the inclination we need the third component of the planetocentric angular momentum, and to compute the latter we need to the minimum unperturbed distance between the particle and the Z-axis. Again, we note that the small body travels on a straight line:

$$x = \frac{U_x}{U_z}z + x_0 \tag{76}$$

$$y = \frac{U_y}{U_z}z. \tag{77}$$

The square of the distance from the Z-axis (not the z-axis!) is:

$$D_Z^2 = x^2 + y^2 = \left(\frac{U_x}{U_z}z + x_0\right)^2 + \left(\frac{U_y}{U_z}z - y_p\right)^2$$

$$= \frac{U_x^2 + U_y^2}{U_z^2}z^2 + \frac{2(U_x x_0 - U_y y_p)}{U_z}z + x_0^2 + y_p^2 \tag{78}$$

and its derivative is:

$$\frac{d(D_Z^2)}{dz} = \frac{2(U_x^2 + U_y^2)}{U_z^2}z + \frac{2(U_x x_0 - U_y y_p)}{U_z}; \tag{79}$$

it is zero at:

$$z = -\frac{(U_x x_0 - U_y y_p)U_z}{U_x^2 + U_y^2}, \tag{80}$$

so that:

$$D_Z^2 = x_0^2 + y_p^2 - \frac{U_x^2 x_0^2 - 2U_x U_y x_0 y_p + U_y^2 y_p^2}{U_x^2 + U_y^2}$$

$$= \frac{(x_0 \cos\theta + y_p \sin\theta \sin\phi)^2}{1 - \sin^2\theta \cos^2\phi}. \tag{81}$$

The third component of the angular momentum is then $D_Z U$, and we have that:

$$\cos i_g = \frac{D_Z U}{bU} = \frac{x_0 \cos\theta + y_p \sin\theta \sin\phi}{b\sqrt{1 - \sin^2\theta \cos^2\phi}}. \tag{82}$$

4.3   The encounter

In general, if the planet is not at the point corresponding to the MOID, but at a generic point $(0, y_p, 0)$, we still have that:

$$\begin{pmatrix} x \\ y \\ z \end{pmatrix} = \begin{pmatrix} U\sin\theta\sin\phi(t - t_0) + x_0 \\ U\cos\theta(t - t_0) \\ U\sin\theta\cos\phi(t - t_0) \end{pmatrix} \tag{83}$$

and we want to minimize the distance from the planet:

$$D^2 = x^2 + (y - y_p)^2 + z^2$$

$$= U^2 t^2 + 2U(x_0 \sin\theta\sin\phi - y_p \cos\theta - U t_0)t$$

$$+ U^2 t_0^2 - 2U(x_0 \sin\theta\sin\phi - y_p \cos\theta)t_0 + x_0^2 + y_p^2 \tag{84}$$

so we take the derivative with respect to t:

$$\frac{d(D^2)}{dt} = 2U^2 t - 2U(y_p \cos\theta - x_0 \sin\theta \sin\phi + Ut_0) \tag{85}$$

and find its zero:

$$t_b = \frac{y_p \cos\theta - x_0 \sin\theta \sin\phi}{U} + t_0. \tag{86}$$

One has the minimum approach distance when the small body is in:

$$\begin{pmatrix} x \\ y \\ z \end{pmatrix} = \begin{pmatrix} \sin\theta \sin\phi(y_p \cos\theta - x_0 \sin\theta \sin\phi) + x_0 \\ \cos\theta(y_p \cos\theta - x_0 \sin\theta \sin\phi) \\ \sin\theta \cos\phi(y_p \cos\theta - x_0 \sin\theta \sin\phi) \end{pmatrix} \tag{87}$$

and the distance is:

$$D \equiv b = \sqrt{x^2 + (y - y_p)^2 + z^2}$$

$$= \sqrt{x_0^2 \cos^2\phi + (x_0 \cos\theta \sin\phi + y_p \sin\theta)^2}. \tag{88}$$

Again we pass from (x, y, z) to the planetocentric axes (X, Y, Z), that differ because in the latter frame the planet is in $(0, 0, 0)$ and not in $(0, y_p, 0)$:

$$\begin{pmatrix} X \\ Y \\ Z \end{pmatrix} = \begin{pmatrix} y_p \sin\theta \cos\theta \sin\phi - x_0 \sin^2\theta \sin^2\phi + x_0 \\ y_p \cos^2\theta - x_0 \sin\theta \cos\theta \sin\phi - y_p \\ y_p \sin\theta \cos\theta \cos\phi - x_0 \sin^2\theta \sin\phi \cos\phi \end{pmatrix} \tag{89}$$

so that the coordinates on the b-plane are:

$$\begin{pmatrix} \xi \\ \eta \\ \zeta \end{pmatrix} = \begin{pmatrix} X \cos\phi - Z \sin\phi \\ (X \sin\phi + Z \cos\phi) \sin\theta + Y \cos\theta \\ (X \sin\phi + Z \cos\phi) \cos\theta - Y \sin\theta \end{pmatrix}$$

$$= \begin{pmatrix} x_0 \cos\phi \\ 0 \\ x_0 \cos\theta \sin\phi + y_p \sin\theta \end{pmatrix}. \tag{90}$$

The expressions for b and ψ are:

$$b \quad = \sqrt{\xi^2 + \zeta^2} = \sqrt{x_0^2 \cos^2\phi + (x_0 \cos\theta \sin\phi + y_p \sin\theta)^2} \tag{91}$$

$$\sin\psi = \frac{\xi}{b} = \frac{x_0 \cos\phi}{\sqrt{x_0^2 \cos^2\phi + (x_0 \cos\theta \sin\phi + y_p \sin\theta)^2}} \tag{92}$$

$$\cos\psi = \frac{\zeta}{b} = \frac{x_0 \cos\theta \sin\phi + y_p \sin\theta}{\sqrt{x_0^2 \cos^2\phi + (x_0 \cos\theta \sin\phi + y_p \sin\theta)^2}} \tag{93}$$

So, given θ and ϕ, the coordinates ξ and ζ on the b-plane depend only on x_0 and y_p; we can invert the relationship and obtain:

$$x_0 = \frac{\xi}{\cos\phi} \tag{94}$$

$$y_p = \frac{\zeta - \cos\theta \tan\phi \xi}{\sin\theta}. \tag{95}$$

At time

$$t_b = \frac{y_p \cos\theta - x_0 \sin\theta \sin\phi}{U} + t_0 = \frac{\cos\theta\zeta - \tan\phi\xi}{U\sin\theta} + t_0, \tag{96}$$

corresponding to the minimum unperturbed distance b, we instantaneously rotate the velocity vector, that is parallel to the incoming asymptote of the planetocentric hyperbola, so as to make it parallel to the other asymptote; moreover, the position of the small body becomes, again instantaneously, the one corresponding to the minimum unperturbed distance on the new orbit; the coordinates in the ξ-η-ζ frame pass from

$$\begin{pmatrix} \xi \\ \eta \\ \zeta \end{pmatrix} = \begin{pmatrix} x_0 \cos\phi \\ 0 \\ x_0 \cos\theta \sin\phi + y_p \sin\theta \end{pmatrix} \tag{97}$$

to

$$\begin{pmatrix} \xi' \\ \eta' \\ \zeta' \end{pmatrix} = \begin{pmatrix} \xi \cos\gamma \\ b \sin\gamma \\ \zeta \cos\gamma \end{pmatrix} \tag{98}$$

and, eliminating γ,

$$\xi' = \frac{(b^2 - c^2)x_0 \cos\phi}{b^2 + c^2} \tag{99}$$

$$\eta' = \frac{2b^2 c}{b^2 + c^2} \tag{100}$$

$$\zeta' = \frac{(b^2 - c^2)(x_0 \cos\theta \sin\phi + y_p \sin\theta)}{b^2 + c^2}. \tag{101}$$

In the X-Y-Z frame

$$\begin{aligned} X' &= (\eta' \sin\theta + \zeta' \cos\theta)\sin\phi + \xi' \cos\phi \\ &= \frac{2b^2 c \sin\theta \sin\phi + (b^2 - c^2)(\cos\theta \sin\phi\zeta + \cos\phi\xi)}{b^2 + c^2} \end{aligned} \tag{102}$$

$$\begin{aligned} Y' &= \eta' \cos\theta - \zeta' \sin\theta \\ &= \frac{2b^2 c \cos\theta - (b^2 - c^2)\sin\theta\zeta}{b^2 + c^2} \end{aligned} \tag{103}$$

$$\begin{aligned} Z' &= (\eta' \sin\theta + \zeta' \cos\theta)\cos\phi - \xi' \sin\phi \\ &= \frac{2b^2 c \sin\theta \cos\phi + (b^2 - c^2)(\cos\theta \cos\phi\zeta - \sin\phi\xi)}{b^2 + c^2}. \end{aligned} \tag{104}$$

We can then easily go back to the x-y-z frame, since

$$\begin{pmatrix} x \\ y \\ z \end{pmatrix} = \begin{pmatrix} X \\ Y - y_p \\ Z \end{pmatrix} \tag{105}$$

so that

$$x'(t_b) = \frac{2b^2 c \sin\theta \sin\phi + (b^2 - c^2)(\cos\theta \sin\phi \zeta + \cos\phi \xi)}{b^2 + c^2} \tag{106}$$

$$y'(t_b) = \frac{2b^2 c \cos\theta - (b^2 - c^2)\sin\theta \zeta}{b^2 + c^2} - \frac{\zeta - \cos\theta \tan\phi \xi}{\sin\theta} \tag{107}$$

$$z'(t_b) = \frac{2b^2 c \sin\theta \cos\phi + (b^2 - c^2)(\cos\theta \cos\phi \zeta - \sin\phi \xi)}{b^2 + c^2}. \tag{108}$$

The components of the new velocity vector are given by

$$\begin{pmatrix} U'_x \\ U'_y \\ U'_z \end{pmatrix} = \begin{pmatrix} U \sin\theta' \sin\phi' \\ U \cos\theta' \\ U \sin\theta' \cos\phi' \end{pmatrix} \tag{109}$$

where θ' and ϕ' are given by the formulae seen before.

The coordinates of the crossing of the ecliptic in the post-encounter branch of the motion can be obtained considering that:

$$\begin{pmatrix} x'(t_b) \\ y'(t_b) \\ z'(t_b) \end{pmatrix} = \begin{pmatrix} U'_x(t_b - t'_0) + x'(t'_0) \\ U'_y(t_b - t'_0) + y'(t'_0) \\ z'(t_b)U'_z(t_b - t'_0) \end{pmatrix}; \tag{110}$$

the time of crossing is then

$$t'_0 = t_b - \frac{z'(t_b)}{U'_z}, \tag{111}$$

the y-coordinate of the crossing is

$$y'(t'_0) \equiv y'_0 = y'(t_b) - U'_y(t_b - t'_0), \tag{112}$$

and the x-coordinate of the crossing is

$$x'(t'_0) \equiv x'_0 = x'(t_b) - U'_x(t_b - t'_0). \tag{113}$$

4.4   The new local MOID

The new local MOID is

$$D'_y = x'_0 \cos\phi'; \tag{114}$$

we can derive it as a function of the pre-encounter b-plane coordinates b and ψ; we begin by rearranging x'_0:

$$\begin{aligned} x'_0 &= x'(t_b) - U'_x(t_b - t'_0) \\ &= b\{\sin\theta \sin\phi \sin\gamma + \cos\theta \sin\phi \cos\psi \cos\gamma + \cos\phi \sin\psi \cos\gamma \\ &\quad - \tan\phi'[\sin\theta \cos\phi \sin\gamma + \cos\theta \cos\phi \cos\psi \cos\gamma \\ &\quad - \sin\phi \sin\psi \cos\gamma]\} \end{aligned} \tag{115}$$

and then compute D'_y:

$$
\begin{aligned}
D'_y &= b\{\cos\phi'[\sin\theta\sin\phi\sin\gamma + \cos\theta\sin\phi\cos\psi\cos\gamma \\
&\quad + \cos\phi\sin\psi\cos\gamma] \\
&\quad - \sin\phi'[\sin\theta\cos\phi\sin\gamma + \cos\theta\cos\phi\cos\psi\cos\gamma \\
&\quad - \sin\phi\sin\psi\cos\gamma]\} \\
&= \frac{b\sin\theta\sin\psi}{\sqrt{(\sin\theta\cos\gamma - \cos\theta\sin\gamma\cos\psi)^2 + \sin^2\gamma\sin^2\psi}}.
\end{aligned}
\tag{116}
$$

Expressing D'_y as a function of b, ξ and ζ:

$$
D'_y = \frac{(b^2 + c^2)\sin\theta\xi}{\sqrt{[(b^2 - c^2)\sin\theta - 2c\cos\theta\zeta]^2 + 4c^2\xi^2}}.
\tag{117}
$$

The implication of the last expression derived for the new local MOID is that, unless $\sin\theta = 0$, i.e. for exactly tangent and coplanar orbits, for which the use of Öpik's theory would be questionable [9], in all practical case the new local MOID cannot be 0 unless the pre-encounter local MOID, i.e. ξ, is not already 0. In other words, initial conditions on the ζ-axis end up on the ζ-axis of the post-encounter b-plane.

4.5   Post-encounter coordinates in the post-encounter b-plane

The coordinates in X-Y-Z frame after the rotation are, as seen before,

$$
X' = \frac{2b^2 c\sin\theta\sin\phi + (b^2 - c^2)(\cos\theta\sin\phi\zeta + \cos\phi\xi)}{b^2 + c^2}
\tag{118}
$$

$$
Y' = \frac{2b^2 c\cos\theta - (b^2 - c^2)\sin\theta\zeta}{b^2 + c^2}
\tag{119}
$$

$$
Z' = \frac{2b^2 c\sin\theta\cos\phi + (b^2 - c^2)(\cos\theta\cos\phi\zeta - \sin\phi\xi)}{b^2 + c^2}.
\tag{120}
$$

We can then apply the appropriate rotations by θ' and ϕ' to obtain the coordinates in the post-encounter b-plane (we denote this reference frame by ξ^*-η^*-ζ^*):

$$
\begin{aligned}
\xi'^* &= X'\cos\phi' - Z'\sin\phi' \\
&= \frac{[2b^2 c\sin\theta\sin\phi + (b^2 - c^2)(\cos\theta\sin\phi\zeta + \cos\phi\xi)]\cos\phi'}{b^2 + c^2} \\
&\quad - \frac{[2b^2 c\sin\theta\cos\phi + (b^2 - c^2)(\cos\theta\cos\phi\zeta - \sin\phi\xi)]\sin\phi'}{b^2 + c^2}
\end{aligned}
\tag{121}
$$

$$
\begin{aligned}
\eta'^* &= (X'\sin\phi' + Z'\cos\phi')\sin\theta' + Y'\cos\theta' \\
&= \Bigg\{ \frac{[2b^2 c\sin\theta\sin\phi + (b^2 - c^2)(\cos\theta\sin\phi\zeta + \cos\phi\xi)]\sin\phi'}{b^2 + c^2}
\end{aligned}
$$

$$+\frac{[2b^2c\sin\theta\cos\phi + (b^2 - c^2)(\cos\theta\cos\phi\zeta - \sin\phi\xi)]\cos\phi'}{b^2 + c^2}\Bigg\}\sin\theta'$$

$$+\frac{[2b^2c\cos\theta - (b^2 - c^2)\sin\theta\zeta]\cos\theta'}{b^2 + c^2} \tag{122}$$

$$\zeta'^* = (X'\sin\phi' + Z'\cos\phi')\cos\theta' - Y'\sin\theta'$$

$$= \Bigg\{\frac{[2b^2c\sin\theta\sin\phi + (b^2 - c^2)(\cos\theta\sin\phi\zeta + \cos\phi\xi)]\sin\phi'}{b^2 + c^2}$$

$$+\frac{[2b^2c\sin\theta\cos\phi + (b^2 - c^2)(\cos\theta\cos\phi\zeta - \sin\phi\xi)]\cos\phi'}{b^2 + c^2}\Bigg\}\cos\theta'$$

$$-\frac{[2b^2c\cos\theta - (b^2 - c^2)\sin\theta\zeta]\sin\theta'}{b^2 + c^2}. \tag{123}$$

A numerical check shows that $\eta'^* = 0$, as expected.

4.6   The next encounter

The orbital period of the planet is 2π, and that of the small body after the encounter is $2\pi a'^{3/2}$; at time

$$t_0'' = t_0' + h \cdot 2\pi a'^{3/2}, \tag{124}$$

where h is an integer, the small body will be again at the same node. On the other hand, in the x-y-z frame (and in the rectilinear motion approximation) the node of the orbit of the small body moves backwards along the y-axis with speed -1; actually, it revolves backwards with period 2π.

At time t_0' the y-component of its distance from the planet was $y_p - y_0'$. We now compute its displacement δy_p along the y-axis between t_0' and t_0'':

$$\delta y_p = -\mathrm{mod}[t_0'' - t_0' + \pi, 2\pi] + \pi. \tag{125}$$

The planetocentric distance of the small body when it is at the node is then

$$D(t_0'') = \sqrt{x_0'^{\,2} + (y_p - y_0' + \delta y_p)^2}. \tag{126}$$

So, in the framework of Öpik's theory, i.e. in terms of simple 2-body heliocentric motion between planetary encounters, the initial conditions for the next encounter are as follows:

- components of the planetocentric velocity: $U_x'' \equiv U_x'$, $U_y'' \equiv U_y'$, $U_z'' \equiv U_z'$;
- time of passage at the node: t_0'';
- nodal distance: $x_0'' \equiv x_0'$;
- distance of the planet from the node: $y_p'' = y_p - y_0' + \delta y_p$.

Let us summarize all the quantities needed to compute the initial conditions of the next encounter, where appropriate as functions of b, ξ and ζ.

The components of the velocity vector $\boldsymbol{U}'$ are computed from:

$$\cos \theta' = \frac{(b^2 - c^2) \cos \theta + 2c \sin \theta \zeta}{b^2 + c^2} \tag{127}$$

$$\sin \theta' = \sqrt{1 - \cos^2 \theta'} \tag{128}$$

$$\cos \phi' = \frac{\cos \phi [(b^2 - c^2) \sin \theta - 2c \cos \theta \zeta] + 2c \sin \phi \xi}{\sqrt{(b^2 - c^2)^2 \sin^2 \theta - 4c[(b^2 - c^2) \sin \theta \cos \theta \zeta - c(b^2 - \sin^2 \theta \zeta^2)]}} \tag{129}$$

$$\sin \phi' = \frac{\sin \phi [(b^2 - c^2) \sin \theta - 2c \cos \theta \zeta] - 2c \cos \phi \xi}{\sqrt{(b^2 - c^2)^2 \sin^2 \theta - 4c[(b^2 - c^2) \sin \theta \cos \theta \zeta - c(b^2 - \sin^2 \theta \zeta^2)]}} \tag{130}$$

$$U_x'' \equiv U_x' = U \sin \theta' \sin \phi' \tag{131}$$

$$U_y'' \equiv U_y' = U \cos \theta' \tag{132}$$

$$U_z'' \equiv U_z' = U \sin \theta' \cos \phi'. \tag{133}$$

The time of passage at the node, near the epoch of the next encounter, is:

$$t_0'' = t_0' + h \cdot 2\pi a'^{3/2}, \tag{134}$$

and the quantities necessary for its computation are:

$$t_0' = t_b - \frac{z'(t_b)}{U_z'} \tag{135}$$

$$t_b = \frac{\cos \theta \zeta - \tan \phi \xi}{U \sin \theta} + t_0 \tag{136}$$

$$z'(t_b) = \frac{2b^2 c \sin \theta \cos \phi + (b^2 - c^2)(\cos \theta \cos \phi \zeta - \sin \phi \xi)}{b^2 + c^2} \tag{137}$$

$$a' = \frac{b^2 + c^2}{(b^2 + c^2)(1 - U^2) - 2U[(b^2 - c^2) \cos \theta + 2c \sin \theta \zeta]}. \tag{138}$$

The nodal distance is:

$$x_0'' = x'(t_b) - U_x'(t_b - t_0'), \tag{139}$$

with:

$$x'(t_b) = \frac{2b^2 c \sin \theta \sin \phi + (b^2 - c^2)(\cos \theta \sin \phi \zeta + \cos \phi \xi)}{b^2 + c^2}. \tag{140}$$

Finally, the new distance of the planet from the node is:

$$y_p'' = y_p - y_0' + \delta y_p, \tag{141}$$

with:

$$y_0' = y'(t_b) - U_y'(t_b - t_0') \tag{142}$$

$$y'(t_b) = \frac{2b^2 c \cos \theta - (b^2 - c^2) \sin \theta \zeta}{b^2 + c^2} - \frac{\zeta - \cos \theta \tan \phi \xi}{\sin \theta} \tag{143}$$

$$\delta y_p = -\mathrm{mod}[t_0'' - t_0' + \pi, 2\pi] + \pi. \tag{144}$$

5   Resonant returns in Öpik's theory

A given resonance corresponds to a certain value of a', i.e. of θ', say a'_0 and θ'_0. If the ratio between the period of the planet and that of the small body is h/k, then:

$$a'_0 = \sqrt[3]{\frac{k^2}{h^2}} \tag{145}$$

$$\cos \theta'_0 = \frac{1 - U^2 - 1/a'_0}{2U} = \frac{1 - U^2 - \sqrt[3]{h^2/k^2}}{2U}. \tag{146}$$

5.1   Solving for a given final semimajor axis

We have

$$\cos \theta'_0 = \cos \theta \cos \gamma + \sin \theta \sin \gamma \cos \psi \tag{147}$$

and, given b, can solve for ψ

$$\cos \psi = \frac{(b^2 + c^2) \cos \theta'_0 - (b^2 - c^2) \cos \theta}{2bc \sin \theta}. \tag{148}$$

Substituting

$$\cos \psi = \frac{\zeta}{b}$$

and

$$b^2 = \xi^2 + \zeta^2,$$

we can solve for ξ as a function of ζ

$$\xi^2 = -\zeta^2 + \frac{2c \sin \theta}{\cos \theta'_0 - \cos \theta} \zeta - \frac{c^2 (\cos \theta'_0 + \cos \theta)}{\cos \theta'_0 - \cos \theta}. \tag{149}$$

This is the equation of a circle centred on the ζ-axis; if R is the radius of such a circle, and D the value of the ζ-coordinate of its centre, its equation is

$$\xi^2 = -\zeta^2 + 2D\zeta + R^2 - D^2; \tag{150}$$

thus, in our case we have a circle centred in

$$D = \frac{c \sin \theta}{\cos \theta'_0 - \cos \theta} \tag{151}$$

and of radius $|R|$, with R given by

$$R = \frac{c \sin \theta'_0}{\cos \theta'_0 - \cos \theta}; \tag{152}$$

note that R is negative for $\theta'_0 > \theta$.

The radius of the circle is zero for $\theta_0' = 0$ and $\theta_0' = \pi$, and for $\theta_0' \to \theta$ both D and R tend to infinity:

$$D = \frac{c \sin \theta}{\cos(\theta + \delta\theta) - \cos \theta} \to -\frac{c}{\delta\theta} \tag{153}$$

$$R = \frac{c \sin(\theta + \delta\theta)}{\cos(\theta + \delta\theta) - \cos \theta} \to -\frac{c}{\delta\theta} - \frac{c \cos \theta}{\sin \theta}. \tag{154}$$

In general, the circle intersects the ζ-axis in

$$\zeta = D \pm R = \frac{c(\sin \theta \pm \sin \theta_0')}{\cos \theta_0' - \cos \theta}, \tag{155}$$

that represent the extremal values that b can take; for $\theta_0' \to \theta$ one of the two intersections tends to infinity, and the other to

$$D - R = -\frac{c}{\delta\theta} + \frac{c}{\delta\theta} + \frac{c \cos \theta}{\sin \theta} = \frac{c \cos \theta}{\sin \theta}. \tag{156}$$

The circle intersects the ξ-axis in

$$\xi = \pm c \sqrt{\frac{(\cos \theta + \cos \theta_0')}{\cos \theta - \cos \theta_0'}}, \tag{157}$$

and the maximum value of $|\xi|$ for which a given θ_0' is accessible is

$$|\xi| \equiv |R| = \left| \frac{c \sin \theta_0'}{\cos \theta_0' - \cos \theta} \right|. \tag{158}$$

The maximum value of a accessible for a given U is for $\theta_0' = 0$, and is obtained for

$$\zeta = \frac{c \sin \theta}{1 - \cos \theta}, \tag{159}$$

and the minimum value of a is for $\theta_0' = \pi$, and is obtained for

$$\zeta = -\frac{c \sin \theta}{1 + \cos \theta}; \tag{160}$$

in both cases for the local MOID we must have $D_y = 0$.

5.2   Examples

Figure 5 shows the circles corresponding to some relevant mean motion resonances on the b-plane of the October 2028 encounter with the Earth of asteroid 1997XF$_{11}$, together with the line denoting a stream of fictitious asteroids all having the same orbital elements and spaced in mean anomaly. The intersections of this line with the circle corresponding to a specific resonance give the region of

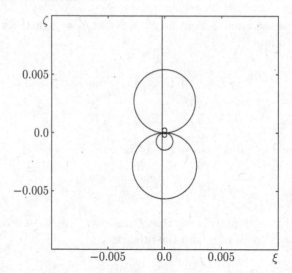

Fig. 5. Circles corresponding to the mean motion resonances (top to bottom) 4/7, 1/2, 2/3, 3/5, 7/12, on the b-plane of the October 2028 encounter with the Earth of asteroid 1997XF$_{11}$. The vertical line represents initial conditions of fictitious asteroids all with the same orbital parameters as 1997XF$_{11}$ and spaced in mean anomaly, i.e. in the time of encounter with the Earth.

the b-plane where the real asteroid has to pass in order to have a 'resonant return', i.e. to be deviated into an orbit of period such that a successive encounter at the same node, due to the resonance relation, will take place [18] [19].

For the same encounter, and the same resonances, Fig. 6 shows the encounter outcomes, computed with Öpik's theory, for a swarm of fictitious asteroids spaced in mean anomaly. Particularly noticeable in the lower right panel is the sharp transition between the final orbits of shortest and longest period.

Figure 7 is the equivalent of Fig. 5 for the b-plane of the August 2027 encounter with the Earth of asteroid 1999AN$_{10}$ and, for the same encounter, Fig. 8 is the equivalent of Fig. 6. In this case U is much larger, so that the effects of the encounter, for the same approach distance, are less noticeable.

6   The distribution of energy perturbations

The heliocentric orbital energy per unit mass of the small body is:

$$E = -\frac{1}{2a};$$
(161)

thus, it is a function only of U and $\cos\theta$:

$$E = \frac{U^2 + 2U\cos\theta - 1}{2};$$
(162)

$\cos\theta$ is then

$$\cos\theta = \frac{1 + 2E - U^2}{2U}$$
(163)

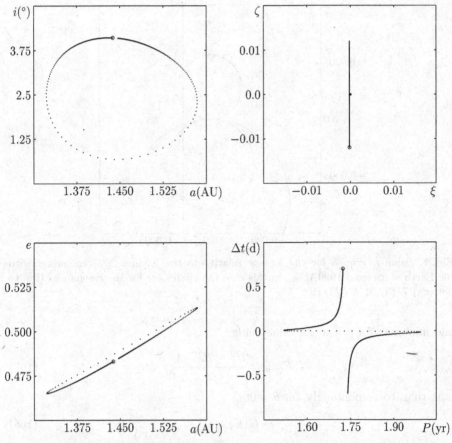

Fig. 6. The outcomes, computed with Öpik's theory, of the October 2028 encounter with the Earth of a swarm of fictitious asteroids with the orbital elements of 1997XF$_{11}$. Upper right: the swarm in the b-plane; upper left: final states in the a-i plane; lower left: final states in the a-e plane; lower right: final states in the plane Δt (difference in time from closest approach; $\Delta t = 0$ for encounter at the MOID) vs P (post-encounter orbital period). In all panels a circle marks one extremum of the swarm, to help identifying the behaviours of the various portions of the swarm.

and $\sin \theta$ is

$$\sin \theta = \sqrt{1 - \left[\frac{1 + 2E - U^2}{2U}\right]^2} = \frac{\sqrt{C(E)}}{2U}, \qquad (164)$$

having put

$$C(E) = -4E^2 - 4(1 - U^2)E - 1 + 6U^2 - U^4. \qquad (165)$$

For a given U, E behaves like a: it is maximum for $\theta = 0$, its value being:

$$E_{max} = \frac{U^2 + 2U - 1}{2} \qquad (166)$$

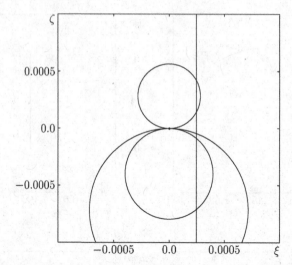

Fig. 7. Same as Fig. 5 for the b-plane relative to the August 2027 encounter with the Earth of asteroid 1999AN$_{10}$. In this case the circles are for the resonances (top to bottom) 7/13, 10/17, 11/19.

and minimum for $\theta = \pi$, its value being:

$$E_{min} = \frac{U^2 - 2U - 1}{2};$$ (167)

note that, correspondingly, for $\theta = 0$

$$C(E) = 0$$ (168)

and for $\theta = \pi$

$$C(E) = 0.$$ (169)

In fact, $C(E)$ can be written as

$$C(E) = -4[E^2 - (E_{max} + E_{min})E + E_{min}E_{max}].$$ (170)

For given U and θ, i.e. for given U and E, the circles in the b-plane leading to a given E_0' are characterized by

$$D = \frac{c\sqrt{C(E)}}{2(E_0' - E)}$$ (171)

$$R = \frac{c\sqrt{C(E_0')}}{2(E_0' - E)}.$$ (172)

Expressing the circles in terms of ξ and ζ, we have:

$$\xi^2 = -\zeta^2 + 2D\zeta + R^2 - D^2$$
$$= -\zeta^2 + \frac{c\sqrt{C(E)}}{E_0' - E}\zeta + \frac{c^2[C(E_0') - C(E)]}{4(E_0' - E)^2}.$$ (173)

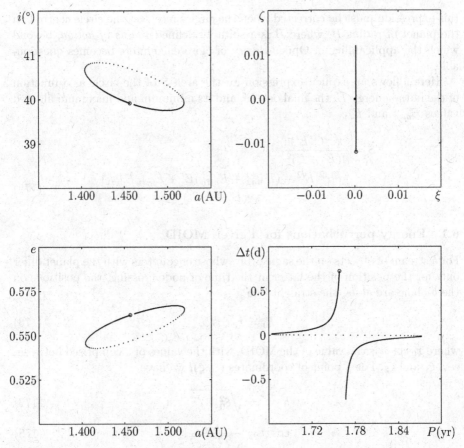

Fig. 8. Same as Fig. 6 for the August 2027 encounter with the Earth of asteroid 1999AN$_{10}$.

It is clear that two circles, relative to two different values E'_1 and E'_2 of the final energy, must be contained one within the other, as they cannot intersect: if they did, the intersection points would lead to two different values of the post-encounter energy, that is clearly impossible. Therefore, the value of the energy perturbation distribution for the interval between E'_1 and E'_2 must be proportional to the difference between the areas of the two relevant circles on the b-plane [25]:

$$\int_{E'_1}^{E'_2} f(E')dE' \propto \frac{c^2 C(E'_1)}{4(E'_1 - E)^2} - \frac{c^2 C(E'_2)}{4(E'_2 - E)^2}$$
$$= -\frac{c^2 [E_1'^2 - (E_{max} + E_{min})E'_1 + E_{min}E_{max}]}{(E'_1 - E)^2}$$
$$+\frac{c^2 [E_2'^2 - (E_{max} + E_{min})E'_2 + E_{min}E_{max}]}{(E'_2 - E)^2}; \quad (174)$$

this expression must be corrected when the circles intersect the circle centred on the planet of radius B, where B is a suitably defined *radius of action*, beyond which the applicability of Öpik's theory of close encounters becomes questionable.

Here follows an explicit expression for the area S of the circle as a function of the initial energy E, the final one E', and its minimum and maximum allowed values E_{min} and E_{max}:

$$
\begin{aligned}
S &= \frac{\pi c^2 C(E')}{4(E' - E)^2} \\
&= -\frac{\pi c^2 [E'^2 - (E_{max} + E_{min})E' + E_{min}E_{max}]}{(E' - E)^2}.
\end{aligned}
\tag{175}
$$

6.1   Energy perturbations for a given MOID

For a stream of objects on the same orbit, whose encounters with the planet differ only for the position of the latter at the time of node crossing, the positions on the b-plane are along the straight line

$$
\xi_0 = x_0 \cos \phi,
\tag{176}
$$

where $x_0 \cos \phi$ is the value of the MOID, with the values of ζ comprised between, say, ζ_1 and ζ_2. For a point of coordinates (ξ_0, ζ), we have

$$
b = \sqrt{\xi_0^2 + \zeta^2}
\tag{177}
$$

$$
\cos \psi = \frac{\zeta}{\sqrt{\xi_0^2 + \zeta^2}},
\tag{178}
$$

so that

$$
\cos \theta' = \frac{\cos \theta \zeta^2 + 2c \sin \theta \zeta + (\xi_0^2 - c^2) \cos \theta}{\zeta^2 + \xi_0^2 + c^2}
\tag{179}
$$

and

$$
E' = E + \frac{2Uc(\sin \theta \zeta - c \cos \theta)}{\zeta^2 + \xi_0^2 + c^2}.
\tag{180}
$$

7   Geocentric variables to characterize meteor orbits

Meteoroid streams are composed of small particles released from comets and from some Earth-crossing asteroids. We see individual meteoroids because their orbits cross that of the Earth, so that they can penetrate in the atmosphere and burn, producing an observable meteor. As these particles are released from their parent bodies at very low relative velocity, initially their orbits do not differ very much from those of their parents. Moreover, because of the Earth-crossing condition, we expect their dynamics to be chaotic.

The problem of identifying meteoroid streams is similar to that of the identification of asteroid families: we have to measure the differences between the orbits of potential family/stream member by a suitably defined distance function, and assess the statistical significance of the groupings obtained in this way.

If the asteroid/meteoroid, after having separated from the parent, were orbiting the Sun in absence of any perturbation, it would continue to do so forever, and its orbital elements would bear a perennial memory of its origin. This is of course not the case in reality, and leads to an important difference between the problems of asteroid families and of meteoroid streams.

In fact:

- asteroid orbits have very little chaoticity, if any, and allow the calculation of 'proper elements', i.e. quasi-integrals of motion stable over Myr to possibly Gyr;
- meteoroid orbits are strongly chaotic, as they cross the orbit of at least one planet (the Earth), and often those of many others.

Therefore, while the timescales over which asteroid families remain recognizable are of the order of the age of the solar system, meteoroid streams can be recognized for much shorter time spans, typically of order $10^3 \div 10^4$ yr.

Figure 9 shows the *radiants*, i.e. the points on the celestial sphere from which the meteor arrives, for 865 precisely measured photographic meteors, in a frame comoving with the Earth, with the Sun on the left. Most orbits are on the upper right of the plot because these are meteors were observed at night in the Northern hemisphere. The concentrations due to some well known meteor streams are easily recognizable.

7.1   An orbital similarity criterion based on geocentric quantities

To classify meteors in streams the standard tool is the orbital similarity criterion originally formulated by Southworth and Hawkins [22], involving the five orbital parameters that describe the orbit (note that q, the perihelion distance, is used instead of a, that is not alway well determined for meteor orbits):

$$D_{SH}^2 = [e_2 - e_1]^2 + [q_2 - q_1]^2 + \left[2\sin\frac{I_{21}}{2}\right]^2$$

$$+ \left[\left(\frac{e_2 + e_1}{2}\right)\left(2\sin\frac{\pi_{21}}{2}\right)\right]^2 \tag{181}$$

where

$$\left[2\sin\frac{I_{21}}{2}\right]^2 = \left[2\sin\frac{i_2 - i_1}{2}\right]^2 + \sin i_1 \sin i_2 \left[2\sin\frac{\Omega_2 - \Omega_1}{2}\right]^2 \tag{182}$$

and

$$\pi_{21} = \omega_2 - \omega_1 + 2\arcsin\left[\cos\frac{i_2 + i_1}{2}\sin\frac{\Omega_2 - \Omega_1}{2}\sec\frac{I_{21}}{2}\right]. \tag{183}$$

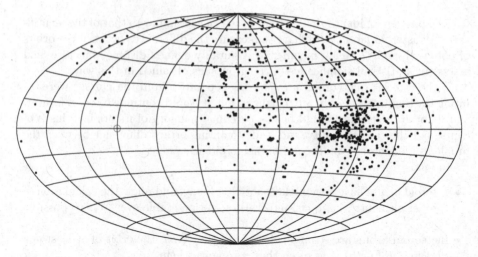

Fig. 9. The radiants of 865 precise photographic meteors in a Hammer-Aitoff equal area projection of the sky, in a system rotating with the Earth about the Sun (here the Earth's orbit is assumed to be circular). The figure is centred in the direction of the motion of the Earth; lines of constant ecliptic latitude B are drawn for $B = 0°, \pm15°, \pm30°, \pm45°, \pm60°, \pm75°$ and of constant ecliptic longitude relative to that of the Sun $L - L_\odot$ for $L - L_\odot = 0°, \pm30°, \pm60°, \pm90°, \pm120°, \pm150°$. The Sun is at $L - L_\odot = -90°, B = 0°$.

When meteors hit the Earth at the ascending node we have

$$\omega + f = 0° \rightarrow f = -\omega, \tag{184}$$

when this happens at the descending node

$$\omega + f = 180° \rightarrow f = 180° - \omega; \tag{185}$$

in both cases, we have that $r = 1$ AU, implying that, at the ascending node,

$$1 = \frac{a(1 - e^2)}{1 + e\cos(-\omega)} \tag{186}$$

and at the descending node

$$1 = \frac{a(1 - e^2)}{1 + e\cos(180° - \omega)}. \tag{187}$$

Therefore, because of the Earth-crossing condition, that involves a, e and ω, a meteor orbit is characterized by only 4 independently measurable quantities. Thus, the identification problem has really only 4 dimensions, while the criterion by Southworth and Hawkins is 5-dimensional; does there exist a suitable set of 4 variables that would allow meteor stream identification in a 'natural' way? And, if so, would not it be better if these variables were directly deducible from observed quantities, without having to derive the orbital elements? Note that the latter, that are necessary to compute D_{SH}, are the conserved quantities of the simplest problem of celestial mechanics, the 2-body problem.

Let us reconsider a meteor fall on the Earth. Its date and time give us Ω of the meteoroid's orbit since, if the the fall takes place at the ascending node, we have

$$\Omega = \lambda_\oplus, \tag{188}$$

and if it takes place at the descending node we have

$$\Omega = \lambda_\oplus + 180°. \tag{189}$$

We need three more quantities to completely characterize the orbit, and we do not want to use a, e, i, ω because they are 4 quantities related by a constraint.

The geometric setup of Öpik's theory suggests us the natural choice: U, θ and ϕ.

What do these variables represent in terms of mentor observables? U is of course the modulus of the geocentric unperturbed velocity, while θ and ϕ define the direction *opposite* to the one from which the meteor is seen to arrive, i.e. opposite to the geocentric radiant. So, U, θ and ϕ are obtained from the directly measurable quantities that characterize an observed meteor.

In Fig. 10 we re-plot the same meteors of Fig. 9, but this time with a grid of given values of θ and ϕ, instead of $L - L_\odot$ and B, so as to allow the reader to get an idea of where orbits of given θ, ϕ are located in the diagram.

Using the same grid, we can plot also the radiants of Apollo asteroids (those with $a > 1$ and $q < 1$), Aten asteroids ($a < 1$) and comets having at least a node between 0.95 and 1.05 AU, so that the similarities and differences between the distribution of meteor radiants, and that of their potential parent bodies, can be appreciated.

Using U, θ and ϕ [24] have defined a new criterion for the similarity of meteor orbits:

$$D_N^2 = [U_2 - U_1]^2 + w_1[\cos\theta_2 - \cos\theta_1]^2 + \Delta\Xi^2 \tag{190}$$

where

$$\Delta\Xi^2 = \min\left[w_2\Delta\phi_I^2 + w_3\Delta\lambda_I^2, w_2\Delta\phi_{II}^2 + w_3\Delta\lambda_{II}^2\right] \tag{191}$$

$$\Delta\phi_I = \left[2\sin\frac{\phi_2 - \phi_1}{2}\right] \tag{192}$$

$$\Delta\phi_{II} = \left[2\sin\frac{180° + \phi_2 - \phi_1}{2}\right] \tag{193}$$

$$\Delta\lambda_I = \left[2\sin\frac{\lambda_2 - \lambda_1}{2}\right] \tag{194}$$

$$\Delta\lambda_{II} = \left[2\sin\frac{180° + \lambda_2 - \lambda_1}{2}\right] \tag{195}$$

and w_1, w_2, w_3 are suitably defined weighting factors; note that $\Delta\Xi$ is small either if both $\phi_1 - \phi_2$ and $\lambda_1 - \lambda_2$ are small, or if they are both close to 180°.

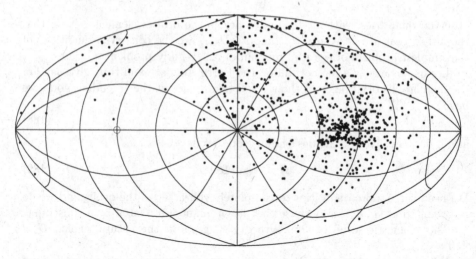

Fig. 10. Same as Fig. 9; the lines here correspond (from the center outwards) to $\theta =$ 150°, 120°, 90°, 60°, 30°, and (from the vertical one in the upper part of the figure, going clockwise) to $\phi =$ 180°, 210°, 240°, 270°, 300°, 330°, 0°, 30°, 60°, 90°, 120°, 150°.

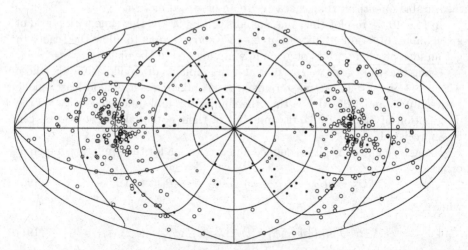

Fig. 11. The radiants of Apollo asteroids (large open circles), Aten asteroids (small open circles) and comets (full dots) having at least a node between 0.95 and 1.05 AU, in the same projection of the sky of Fig. 10. Note that close to the centre of the figure there are only comets and Atens, while Apollos avoid that region and fill up, together with comets, all the rest of the diagram. There are noteworthy concentrations of Apollo radiants in the directions towards and away from the Sun.

This allows to recognize as members of the same stream meteors observed at node crossings separated by 6 months (e.g. the Orionids and the η Aquarids).

The effectiveness of D_N for the identification of meteoroid streams was tested by applying it to a suitable set of meteor data, and comparing its results to those obtained with D_{SH} [14]. The data used comprised 865 precise orbits taken from the IAU Meteor Data Center files; these orbits are of 139 small-camera meteors [27], of 413 Super Schmidt meteors [12], and of 313 Super Schmidt meteors [10] [11]. The groupings were obtained by a single neighbour linking algorithm [17] not requiring any a priori knowledge of stream orbits, and the thresholds for D_{SH} and D_N were determined by comparison with random samples having the same marginal distributions of the variables [13], in order to have a reliability level of 99%.

The comparison between the classifications made with D_{SH} and D_N gave the following results [14]:

- 15 streams were identified both by D_{SH} and by D_N;
- 6 streams (κ Cygnids, Quadrantids, Geminids, Southern δ Aquarids, Draconids and Cyclids) showed identical memberships in the two cases;
- for the Orionids the membership was also identical, but D_N failed to identify as belonging to the same stream an η Aquarid meteor;
- for 7 streams (Lyrids, α Capricornids, Perseids, Taurids, Leonids, σ Hydrids and α Pegasids) D_N was able to add a few more members, and for the χ Orionids D_N was able to add many more members, to those identified by D_{SH};
- two streams were identified only with D_{SH} (σ Leonids and Andromedids);
- five streams were identified only with D_N (ε Geminids, Monocerotids, α Virginids, Northern δ Aquarids and ε Piscids);
- D_N detected a single Southern α Capricornid, and the Northern branch of the χ Orionids, both missed by D_{SH};
- of the streams identified only by D_N two, the α Virginids and the ε Piscids, are near-ecliptical and possess a Northern and a Southern branch;
- the two streams identified by D_{SH} and not by D_N possess a Northern and a Southern branch.

Thus, D_{SH} and D_N gave essentially equivalent classifications for streams of moderate to high inclination, while for near-ecliptical streams the memberships disagreed significantly in some cases.

7.2   Secular invariance of U and θ

Many factors influence the dynamical evolution of meteoroid streams, and some of them are due to forces other than gravitation. However, over not too long timescales, and in absence of planetary close encounters, we can assume that only secular perturbations affect meteoroid orbits. The most important secular perturbation in the meteoroid stream case is the one related to the cycle of ω, first described by Kozai [16]. Assuming that all the planets are on circular coplanar orbits, and that the small body orbit is far from mean motion and secular resonances, it leaves invariant the z-component of the orbital angular momentum:

$$L_z = \sqrt{a(1 - e^2)} \cos i \qquad (196)$$

and the orbital energy:

$$E = -\frac{1}{2a}; \qquad (197)$$

therefore,

$$T = \frac{1}{a} + 2\sqrt{a(1 - e^2)} \cos i = 2(L_z - E) \qquad (198)$$

is constant, and so is U. But then, if a and U are conserved, so is θ, since:

$$\cos \theta = \frac{1 - U^2 - 1/a}{2U}. \qquad (199)$$

The secular invariance of U and $\cos \theta$ suggests that they can be of some use to identify, as possibly being originated from the same parent body, streams that are on orbits of different e, i, ω and Ω, due to secular perturbations [14].

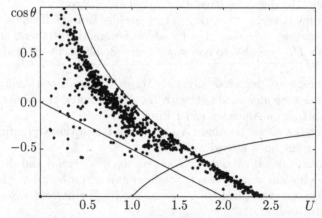

Fig. 12. The positions of 865 precisely measured photographic meteor orbits in the plane U-$\cos \theta$. The straight line going from $(0,0)$ to $(2,-1)$ is the locus of orbits with $a = 1$ AU, while the curve from the upper left to the lower right is the locus of orbits with $a = \inf$; the remaining curve is the locus of orbits with $L_z = 0$.

In fact, if we compare the positions of photographic meteor orbits in the plane U-$\cos \theta$ (Fig. 12) with those of known near-Earth asteroids and comets in the same plane (Fig. 13), we can immediately notice that, except for some overlapping in a limited region of the diagram, asteroids and comets tend to reside in different regions of this diagram. In practice this means that the knowledge of just U and θ of a meteor orbit, something immediately deducible from observations, without the need to compute the orbital elements, allows in most cases to guess whether the parent body is on an asteroidal or a cometary orbit.

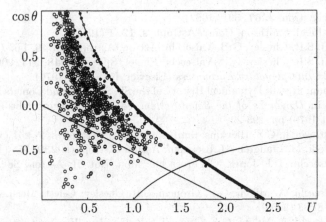

Fig. 13. The positions of Apollo, Aten and Amor asteroids (open circles) and comets (dots) in the plane U-$\cos\theta$. Note that comes and asteroids overlap only on a limited region of the diagram.

Acknowledgements

I am indebted to the people with whom I have worked on this subject over many years: Andrea Carusi, Richard Greenberg, Claude Froeschlé, Tadeusz Jopek, Andrea Milani, Steve Chesley, Giovanni Gronchi.

References

1. L.A.M. Benner, W.B. McKinnon: Icarus **118**, 155 (1995)
2. C. Bonanno: Astron. Astrophys. **360**, 411 (2000)
3. .A. Carusi, G.B. Valsecchi: 'Numerical Simulations of Close Encounters between Jupiter and Minor Bodies'. In *Asteroids*, ed. by T. Gehrels (Univ. Arizona Press, Tucson 1979) pp. 391-416
4. A. Carusi, G.B. Valsecchi: Astron. Astrophys. **94**, 226 (1981)
5. A. Carusi, Ľ. Kresák, G.B. Valsecchi: Astron. Astrophys. **99**, 262 (1981)
6. A. Carusi, Ľ. Kresák, E. Perozzi, G.B. Valsecchi: *Long-Term Evolution of Short-Period Comets* (Adam Hilger, Bristol 1985)
7. A. Carusi, G.B. Valsecchi, R. Greenberg: Celest. Mech. & Dynam. Astron. **49**, 111 (1990)
8. A. Carusi, Ľ. Kresák, G.B. Valsecchi: *Electronic Atlas of Dynamical Evolutions of Short-Period Comets* (available on the World Wide Web at the URL http://www.ias.rm.cnr.it/ias-home/comet/catalog.html 1995)
9. R. Greenberg, A. Carusi, G.B. Valsecchi: Icarus **75**, 1 (1988)
10. G.S. Hawkins, R.B. Southworth: Smithson. Contr. Astrophys. **2**, 349 (1958)
11. G.S. Hawkins, R.B. Southworth: Smithson. Contr. Astrophys. **4**, 85 (1961)
12. L.G. Jacchia, F.L. Whipple: Smithson. Contr. Astrophys. **4**, 97 (1961)
13. T.J. Jopek, Cl. Froeschlé: Astron. Astrophys. **320**, 631 (1997)
14. T.J. Jopek, G.B. Valsecchi, Cl. Froeschlé: Mon. Not. R. Astron. Soc. **304**, 751 (1999)
15. D.M. Kary, L. Dones: Icarus **121**, 207 (1996)

16. Y. Kozai: Astron. J. **67**, 591 (1962)
17. B.A. Lindblad: Smithson. Contr. Astrophys. **12**, 1 (1971)
18. A. Milani, S.R. Chesley, G.B. Valsecchi: Astron. Astrophys. **346**, L65 (1999)
19. A. Milani, S.R. Chesley, G.B. Valsecchi: Planet. Space Sci. **48**, 945 (2000)
20. E.J. Öpik: *Interplanetary Encounters* (Elsevier, New York 1976)
21. H. Rickman: 'Recent Dynamical History of the Six Short-Period Comets Discovered in 1975'. In *Dynamics of the Solar System*, ed. by R.L. Duncombe (D. Reidel, Dordrecht 1979) pp. 293-298
22. R.B. Southworth, G.S. Hawkins: Smithson. Contr. Astrophys. **7**, 261 (1963)
23. G. Tancredi, M. Lindgren, H. Rickman: Astron. Astrophys. **239**, 375 (1990)
24. G.B. Valsecchi, T.J. Jopek, Cl. Froeschlé: Mon. Not. R. Astron. Soc. **304**, 743 (1999)
25. G.B. Valsecchi, A. Milani, G.F. Gronchi, S.R. Chesley: Celest. Mech. & Dynam. Astron. **78**, 83 (2000)
26. G.B. Valsecchi, A. Milani, G.F. Gronchi, S.R. Chesley: 'Resonant returns to close approaches: analytical theory', submitted to Astron. Astrophys. (2001)
27. F.L. Whipple: Astron. J. **59**, 201 (1954)

Generalized Averaging Principle and Proper Elements for NEAs

Giovanni-Federico Gronchi

Dipartimento di Matematica, Università di Pisa,
Via Buonarroti 2, 56127 Pisa, Italy

Abstract. We present a review of the results concerning a generalization of the classical averaging principle suitable to deal with orbit crossings, that make singular the Newtonian potential at the values of the anomalies corresponding to collisions. These methods have been applied to study the secular evolution of Near Earth Asteroids and to define proper elements for them, that are useful to study the possibility of impact between these asteroids and the Earth.

1 Introduction

The averaging principle is a powerful tool to study the qualitative behavior of the solutions of Ordinary Differential Equations. It consists in solving averaged equations, obtained by an integral average of the original equations over some angular variables; if some conditions are satisfied the solutions of the averaged equations remain close to the solutions of the original equations for a long time span. A review of the classical results on averaging methods in perturbation theory can be found in [1].

These methods have been used to study the secular evolution of the Main Belt Asteroids (MBAs) starting from [26], see [11],[22],[15] .

On the other hand in the case of Near Earth Asteroids (NEAs) the intersections between the orbits of the asteroid and those of the planets generate singularities in the Newtonian potential corresponding to the collision values of the phases on their orbits: in this case the averaged equations have no meaning.

In 1998 Gronchi and Milani [8] have defined piecewise differentiable solutions that can be regarded as solutions of the averaged equations in a weak sense: they solve slightly modified averaged equations in which an inversion of the integral and differential operators occurs. These equations correspond to the classical averaged equations when there are no crossings between the orbits.

The eccentricity and the inclination of the Solar System planets are not considered in this framework: this simplification gives rise to some nice properties, like the periodicity of the solutions of the averaged equations with respect to the perihelion argument, and it allows to prove a stability property [9].

Using the generalized averaging principle Gronchi and Milani [10] computed proper elements and proper frequencies for all the known NEAs using the NEODyS database of orbits (http://newton.dm.unipi.it/neodys/). The related catalog is continuously updated according to the discovery of new asteroids and

to the changes in the orbits of the known ones: it can be found at the web address
http://newton.dm.unipi.it/neodys/propneo/catalog.tot .

The reliability of these solutions has been tested by a comparison with the outputs of pure numerical integrations [7] and the results are quite satisfactory.

Recently this generalized averaging theory has been extended to the eccentric/inclined case for the planets [5]; this will allow to define more reliable crossing times between the orbits, that are useful to detect the possibility of collisions.

In this paper we shall review the classical averaging principle for non crossing orbits and we shall describe in all details the generalization of the principle when a crossing occurs in the case with the planets on circular coplanar orbits. Then we shall present an application of the generalized principle to compute the secular evolution of NEAs and proper elements for them. Finally two short sections are devoted to discussions on the reliability of the averaged orbits and to the recent work that extends the averaging theory including the eccentricity and inclination of the planets.

2   The classical averaging principle

First we shall write canonical equations of motion to compute the time evolution of the orbit of an asteroid. Then we shall describe the averaged equations for the evolution of asteroids that do not cross the orbits of the planets.

2.1   The full equations of motion

Let us consider a Solar System model with the Sun, $N - 2$ planets and an asteroid: we assume that the mass of the asteroid is negligible, so that we have a *restricted problem*. We also suppose that the masses of the planets are small if compared with the mass of the Sun, so that we have $N - 2$ small perturbative parameters $\mu_i, i = 1 \ldots N - 2$, corresponding to the ratio of the mass of each planet with the mass of the Sun.

We assume that the motion of the planets is completely determined and that there are no collisions among them or with the Sun. With these assumptions we write the full equations of motion for the asteroid in Hamiltonian form.

We use heliocentric Delaunay's variables for the asteroid, defined by

$$
\begin{cases}
L = k\sqrt{a} \\
G = k\sqrt{a(1 - e^2)} \\
Z = k\sqrt{a(1 - e^2)} \cos I
\end{cases}
\qquad
\begin{cases}
\ell = n(t - t_0) \\
g = \omega \\
z = \Omega
\end{cases}
$$

where $\{a, e, I, \omega, \Omega, \ell\}$ is the set of the Keplerian elements, k is Gauss's constant, n is the mean motion and t_0 is the time of passage at perihelion.

Delaunay's variables , like the Keplerian elements, describe the evolution of the osculating orbit of the asteroid, that is of the trajectory that the asteroid would describe in a heliocentric reference frame, given its position and velocity at a time t, if only the Sun were present. For negative values of the Keplerian energy of the asteroid the osculating orbits are ellipses; we shall consider only such cases.

The Hamiltonian can be written as

$$H = -\frac{k^2}{2L^2} - R$$

where $-k^2/(2L^2)$ is the *unperturbed term*, describing the two body motion of the asteroid around the Sun, and R is the *perturbing function* defined by

$$R = \sum_{i=1}^{N-2} \mu_i R_i; \quad R_i = k^2 \left[\frac{1}{|x - x_i|} - \frac{<x, x_i>}{|x_i|^3} \right]; \quad i = 1 \ldots N - 2 \quad (1)$$

in which $\langle , \rangle$ is the Euclidean scalar product and x and x_i are the position vectors of the asteroid and of all the planets in a heliocentric reference frame.

Note that each R_i is the sum of a *direct term* $k^2/|x - x_i|$, due to the direct interaction between the planet i and the asteroid, and an *indirect term* $-k^2 < x, x_i > /|x_i|^3$, representing the effects on the motion of the asteroid caused by the interaction between the Sun and the planet i.

If we set $\mathfrak{E}_D = (L, G, Z, \ell, g, z)$ we can write Hamilton's equations as

$$\dot{\mathfrak{E}}_D = \mathfrak{J} (\nabla_{\mathfrak{E}_D} H)^t \quad (2)$$

where the *dot* means derivative with respect to time, $\mathfrak{J}$ is the 6×6 matrix

$$\begin{bmatrix} \mathcal{O} & -\mathcal{I}_3 \\ \mathcal{I}_3 & \mathcal{O} \end{bmatrix}$$

composed by 3×3 zero and identity matrixes, and

$$(\nabla_{\mathfrak{E}_D} H)^t = \left(\frac{\partial H}{\partial \mathfrak{E}_D} \right)^t$$

is the transposed vector of the partial derivatives of the Hamiltonian H with respect to $\mathfrak{E}_D$.

2.2   The averaged equations

The classical averaging principle consists in solving equations obtained by the integral average of the right hand side of (2) over the mean anomalies $\ell, \ell_1, .., \ell_{N-2}$ of the asteroid and the planets.

This method can be applied to study the qualitative behavior of the orbits of the MBAs, that do not cross the orbits of the Solar System planets during their evolution, assuming that *no mean motion resonances with low order occur* between the asteroid and the planets in the model. This means that there exists $\epsilon > 0$ not too small and a positive integer M not too large such that for each pair $(a(t), a_i(t))$, composed by the semimajor axes of the osculating orbits of the asteroid and the planet i $(i = 1 \ldots N - 2)$ we have

$$\left| p\, [a(t)]^{3/2} - q\, [a_i(t)]^{3/2} \right| > \epsilon$$

for each pair of positive integers $p, q \leq M$ and for each t in the considered time span.

Remark 1. As our purpose is not a study of the structure of the mean motion resonances we shall not give further details or any estimates on the size of ϵ and M.

In the expression of the perturbing function (1) the effect of each planet is independently taken into account: each R_i is a function of the coordinates and the masses of the asteroid and one planet only. We shall study the case of only one perturbing planet and we shall use a prime for the quantities related to this planet: the perturbation of all the planets, up to the first order in the perturbing masses μ_i, will be obtained by the sum of the contribution of each planet.

If we consider the reduced set of Delaunay's variables $E_D = (G, Z, g, z)$ the averaged equations for the asteroid can be written in the following form:

$$\dot{\widetilde{E}}_D = -\mathcal{J}\, \overline{\nabla_{E_D} R}^{\, t} \tag{3}$$

where $\widetilde{E}_D = (\tilde{G}, \tilde{Z}, \tilde{g}, \tilde{z})$ are averaged Delaunay's variables, $\mathcal{J}$ is the 4×4 matrix

$$\begin{bmatrix} \mathcal{O} & -\mathcal{I}_2 \\ \mathcal{I}_2 & \mathcal{O} \end{bmatrix}$$

composed by 2×2 zero and identity matrixes, and $\overline{\nabla_{E_D} R}^{\, t}$ is the transposed vector of the integral average over (ℓ, ℓ') of the partial derivatives of the perturbing function R with respect to E_D

$$\overline{\nabla_{E_D} R} = \frac{1}{(2\pi)^2} \int_{-\pi}^{\pi} \int_{-\pi}^{\pi} \nabla_{E_D} R \, d\ell \, d\ell' \; ; \qquad \nabla_{E_D} R = \frac{\partial R}{\partial E_D} \, .$$

Remark 2. As we are considering non–crossing orbits, the derivatives of R with respect to Delaunay's variables are regular functions and we can use the *theorem of differentiation under the integral sign* [3] to exchange the derivatives and the integrals in (3); then the averaged equations take the form

$$\dot{\widetilde{E}}_D = -\mathcal{J}\, (\nabla_{E_D} \overline{R})^{\, t} \tag{4}$$

where

$$\overline{R} = \frac{1}{(2\pi)^2} \int_{-\pi}^{\pi} \int_{-\pi}^{\pi} R \, d\ell \, d\ell' = \frac{1}{(2\pi)^2} \int_{-\pi}^{\pi} \int_{-\pi}^{\pi} \frac{\mu k^2}{|x - x'|} \, d\ell \, d\ell' \tag{5}$$

(μ is the ratio between the mass of the planet and the mass of the Sun) because the average of the indirect term of the perturbing function is zero (see [26]).

Remark 3. We shall skip the 'tilde' over the averaged variables in the following to avoid the use of heavy notations.

We stress that the solutions of (3) are representative of the solutions of the full equations of motion only if there are no mean motion resonances with low order between the asteroid and the planet.

2.3 Difficulties arising with crossing orbits

We say that an asteroid is *planet crossing* if its orbit crosses the orbit of some planet during its secular evolution.

When we consider a planet crossing asteroid at the time of intersection of the orbits, the averaged perturbing function $\overline{R}$ is the integral of an unbounded function that is convergent because $1/|x - x'|$ has a first order polar singularity in the values $\overline{\ell}, \overline{\ell}'$ corresponding to a collision. The derivatives at the right hand side of (3) have second order polar singularities in $\overline{\ell}, \overline{\ell}'$, hence equations (3) do not make sense in this case because the integrals over ℓ, ℓ' of these derivatives are divergent and *the classical averaging principle cannot be applied.*

3   Generalized averaging principle in the circular coplanar case

We present the ideas of the generalization of the averaging principle to the case of crossing orbits, assuming that all the planets in the model have *circular and coplanar orbits* (see [13]) and that no low order mean motion resonances are present.

The natural choice for a heliocentric reference frame is then a system $Oxyz$ with the (x, y)-plane corresponding to the common orbital plane of all the planets, oriented in such a way that the planets have positive z component of the angular momentum with respect to the origin O.

3.1   Geometry of the node crossing

We assume that *the inclination between the osculating orbit of the asteroid with respect to the orbital plane of the planets is different from zero during its whole evolution*; then it is possible to define, for all times, the *mutual nodal line*, representing the intersection of the two orbital planes of the asteroid and the planets.

Let us consider one planet at a time: we give the following

Definition 1. We call *mutual node* each pair of points on the mutual nodal line, one belonging to the orbit of the asteroid and the other to the one of the planet, that lie on the same side of the mutual nodal line with respect to the common focus of the two conics. For each planet in this model there are two mutual nodes, the *ascending* and the *descending* one: they differ in the change of sign of the z component along the asteroid orbit (negative to positive in the first case and vice-versa in the second).

We say that an *ascending (resp. descending) node crossing* occurs when the orbit of the asteroid intersects the orbit of a planet at the ascending (resp. descending) mutual node, that in this case becomes a set of two coinciding points.

Unless the inclination of the asteroid never vanishes, the only way to have orbital intersection is a node crossing. In the following we give a description of the possible geometric configurations of node crossings in the plane $(e \cos \omega, e \sin \omega)$.

Recall that an ascending and descending node crossing with a planet whose orbit has semimajor axis a'_i is characterized by the vanishing of the following expressions respectively:

$$d^+_{nod}(i) = \frac{a(1-e^2)}{1+e\cos\omega} - a'_i \ ; \qquad d^-_{nod}(i) = \frac{a(1-e^2)}{1-e\cos\omega} - a'_i \qquad (6)$$

that are called *nodal distances* (and can be negative).

In the averaged problem with the planets on circular coplanar orbits we have three integrals of motions: the semimajor axis a, the *Kozai integral* $\overline{H} = H_0 - \overline{R}$ (that is the averaged Hamiltonian) and the *z-component of the angular momentum* $Z = k\sqrt{a(1-e^2)}\cos I$. The Z integral allows to determine the evolution of $I(t)$ if we know $e(t)$; if we also know $\omega(t)$ we can determine $\Omega(t)$ by a simple quadrature of $\partial \overline{R}/\partial Z$, that does not depend on Ω. From the expression of the integral Z we deduce the maximum value of the averaged inclination and eccentricity:

$$I_{max} = I\big|_{e=0} = \arccos \frac{Z}{k\sqrt{a}} \ ; \qquad e_{max} = e\big|_{I=0} = \frac{\sqrt{k^2a - Z^2}}{k\sqrt{a}} \ .$$

For a given value of the semimajor axis a we can represent the level lines of the averaged Hamiltonian, on which the averaged solutions evolve, in the plane $(\xi, \eta) := (e \cos \omega, e \sin \omega)$. We define the *Kozai domain*

$$W = \{(\xi, \eta) : \xi^2 + \eta^2 \le e^2_{max}\},$$

where the averaged dynamics is confined.

In the (ξ, η) reference plane the node crossing lines with the planets are circles: they are defined by

$$\Gamma^+(i) = \{(\xi, \eta) : d^+_{nod}(i) = 0\} \ ; \qquad \Gamma^-(i) = \{(\xi, \eta) : d^-_{nod}(i) = 0\}$$

where i is the index of the planet.

At the ascending node crossing with the planet i the equation to be considered is

$$1 - \xi^2 - \eta^2 = \frac{a'_i}{a}(1 + \xi) \ .$$

After the coordinate change $X = \xi + a'_i/(2a); \ Y = \eta$ we obtain

$$X^2 + Y^2 = \left(1 - \frac{a'_i}{2a}\right)^2,$$

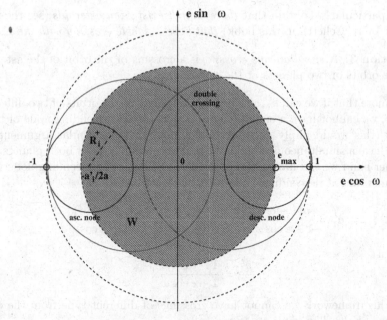

Fig. 1. The Kozai domain $\{(e\cos\omega, e\sin\omega) : 0 \leq e \leq e_{max}, \omega \in \mathbb{R}\}$ is represented in the figure by the set W. We plot also the four circles corresponding to ascending and descending node crossing with two planets (they have their centers shifted respectively on the left and on the right). An additional exterior circle corresponding to the boundary for closed orbits $(e = 1)$ is drawn.

that is, in the (ξ, η)-plane, the equation of a circle of radius $R_i^+ = 1 - a_i'/(2a)$, with center in $(\xi_+, \eta_+) = (-a_i'/(2a), 0)$ (see Fig. 1).

By the previous calculations we have

$$\begin{cases} d_{nod}^+(i) > 0 \quad \text{inside } \Gamma^+(i) \\[2mm] d_{nod}^+(i) < 0 \quad \text{outside } \Gamma^+(i) \ . \end{cases}$$

In a similar way we can prove that the equation $d_{nod}^-(i) = 0$ represents a circle of radius $R_i^- = R_i^+ = 1 - a_i'/(2a)$, with center in $(\xi_-, \eta_-) = (+a_i'/(2a), 0)$.

Definition 2. A *double (node) crossing* is a crossing between the orbit of the asteroid and the orbit of a planet at both the ascending and descending node.

By the symmetry of the circles $\{d_{nod}^+(i) = 0\}$ and $\{d_{nod}^-(i) = 0\}$, for each index i, we can deduce that a double crossing is possible only when $\omega = \pi/2$ or $\omega = 3\pi/2$ (see Fig. 1). We obtain the following condition on the ratio of the semimajor axes a, a_i':

$$-\frac{a_i'}{2a} \geq -\frac{1}{2} \qquad \text{that is} \qquad a \geq a_i' , \tag{7}$$

and in particular we obtain that *there cannot exist Aten asteroids* (see the first chapter by A. Celletti, in this book) *that have a double crossing with the Earth.*

Definition 3. A *simultaneous crossing* is a crossing of the orbit of the asteroid and the orbits of two planets at the same time.

We note that if we call a'_1, a'_2 the semimajor axes of the orbits of two different planets, we cannot have a simultaneous crossing at the ascending node of both planets (this would imply $a'_1 = a'_2$ in this model). By a similar argument we cannot have a simultaneous crossing at the descending node of both planets. On the other hand we can have a simultaneous crossing at the ascending node with one planet and at descending node with the other one if

$$a'_1 = \frac{a(1 - e^2)}{1 + e \cos \omega} \qquad \text{and} \qquad a'_2 = \frac{a(1 - e^2)}{1 - e \cos \omega} \;,$$

that is if

$$\frac{a'_1}{a'_2} = \frac{1 - e \cos \omega}{1 + e \cos \omega} \;.$$

In this framework we cannot have crossings of different type from the ones presented above (like *triple crossing*, etc.).

3.2   Description of the osculating orbits

We consider a model with three bodies only: Sun, planet, asteroid. We set the x axis along the line of the nodes, pointing towards the ascending mutual node. The equations defining the osculating orbits $P(u) = (p_1(u), p_2(u), p_3(u))$ and $P'(u') = (p'_1(u'), p'_2(u'), p'_3(u'))$ of the asteroid and the planet are

$$\begin{cases} p_1 = a[(\cos u - e) \cos \omega - \beta \sin u \sin \omega] \\ p_2 = a[(\cos u - e) \sin \omega + \beta \sin u \cos \omega] \cos I \\ p_3 = a[(\cos u - e) \sin \omega + \beta \sin u \cos \omega] \sin I \end{cases} \qquad \begin{cases} p'_1 = a' \cos u' \\ p'_2 = a' \sin u' \\ p'_3 = 0 \end{cases} \quad (8)$$

where u, u' are the eccentric anomalies and $\beta = \sqrt{1 - e^2}$. These orbits are respectively an ellipse and a circle.

The distance between a point on an orbit and a point on the other one, appearing at the denominator of the direct term of the perturbing function, is defined by its square as

$$\mathfrak{D}^2(u, u') = (p_1 - p'_1)^2 + (p_2 - p'_2)^2 + (p_3 - p'_3)^2 =$$
$$a^2(1 - e \cos u)^2 + a'^2 - 2aa' \big\{ \cos u'[(\cos u - e) \cos \omega - \beta \sin u \sin \omega] +$$
$$+ \sin u' \cos I[(\cos u - e) \sin \omega + \beta \sin u \cos \omega] \big\} \;.$$

We introduce the function $D(\ell, \ell')$, which is implicitly defined by

$$D(\ell(u), \ell'(u')) = \mathfrak{D}(u, u') \tag{9}$$

and by Kepler's equations

$$\ell = u - e \sin u; \qquad\qquad \ell' = u' \qquad\qquad (10)$$

for the asteroid and the planet (the latter has a simpler form because the orbit of the planet is circular).

We define the values of the anomalies $\overline{u}, \overline{u}'$ corresponding to the mutual ascending node: we immediately notice that $\overline{u}' = 0$, while from

$$a(1 - e \cos \overline{u}) = \frac{a\beta^2}{1 + e \cos \omega} \qquad\qquad (11)$$

we obtain

$$\cos \overline{u} = \frac{\cos \omega + e}{(1 + e \cos \omega)}; \qquad \sin \overline{u} = -\frac{\beta \sin \omega}{(1 + e \cos \omega)}$$

(the sign of $\sin \overline{u}$ has been chosen in such a way that it is opposite to the sign of $\sin \omega$).

The equations defining the anomalies $\overline{u}_1, \overline{u}'_1$, corresponding to the mutual descending node, are

$$\overline{u}'_1 = \pi; \qquad \cos \overline{u}_1 = \frac{e - \cos \omega}{(1 - e \cos \omega)}; \qquad \sin \overline{u}_1 = \frac{\beta \sin \omega}{(1 - e \cos \omega)}.$$

In the following we shall study only ascending node crossings, but the same methods are suitable to deal also with the descending ones and even with double crossings (see [6],[10]).

3.3 Weak averaged solutions

The idea of the generalization of the averaging principle comes from Remark 2: if there are no crossings between the orbits, then the averaged equations of motion (3) are equivalent to equations (4).

We write equations (4) in a more explicit form:

$$\begin{cases} \dot{\tilde{G}} = \dfrac{\partial \overline{R}}{\partial g} \\[2mm] \dot{\tilde{Z}} = \dfrac{\partial \overline{R}}{\partial z} = 0 \end{cases} \qquad \begin{cases} \dot{\tilde{g}} = -\dfrac{\partial \overline{R}}{\partial G} \\[2mm] \dot{\tilde{z}} = -\dfrac{\partial \overline{R}}{\partial Z} ; \end{cases} \qquad (12)$$

the equation $\dot{\tilde{Z}} = 0$ holds because the equations of the orbits do not depend on the longitude of the node Ω.

We shall prove that when the orbits intersect each other it is possible to define piecewise smooth solutions of equations (12), that we call *weak averaged solutions*, and we shall see that the loss of regularity corresponds exactly to the crossing configurations of the orbits: in fact we shall give a twofold meaning to the right hand sides of (12) at the node crossing, corresponding to the two limit values of the derivatives coming from inside and outside the circle representing the ascending node crossing with the planet in the plane (ξ, η).

Note that the weak averaged solutions correspond to the classical averaged solutions as far as their trajectories in the reduced phase space (ξ, η) do not pass through a node crossing line.

We also observe that the exchange of the differential and integral operators in (12) is not essential for a theoretical definition of the weak solutions (they could anyway be defined as the limits of the solutions of (3) coming from both sides of the node crossing lines) but, as we shall see, this operation is necessary to obtain analytic formulas for the discontinuity of the average of the derivatives of R, that are not defined on the node crossing lines, and to define the semianalytic procedure to compute the weak solutions.

3.4   The Wetherill function

Let $\{P(\overline{u}), P'(\overline{u}')\}$ be the ascending mutual node. We consider the two straight lines $r(\ell)$ and $r'(\ell')$, tangent in $P(\overline{u})$ and $P'(\overline{u}')$ to the orbits of the asteroid and of the planet (see Fig. 2); they can be parametrized by the mean anomalies ℓ, ℓ' so that $P(u(t))$ and $r(\ell(t))$ have the same velocities (derivatives with respect to t) in $P(\overline{u})$ and $P'(u'(t))$ and $r'(\ell'(t))$ have the same velocities in $P'(\overline{u}')$:

$$
\begin{cases} r_1 = \overline{x} - \mathcal{F}(\ell - \overline{\ell}) \\ r_2 = \overline{y} + \mathcal{G}\cos I(\ell - \overline{\ell}) \\ r_3 = \overline{z} + \mathcal{G}\sin I(\ell - \overline{\ell}) \end{cases}
\qquad
\begin{cases} r_1' = \overline{x}' \\ r_2' = \overline{y}' + a'(\ell' - \overline{\ell}') \\ r_3' = \overline{z}' \end{cases}
\qquad (13)
$$

where $\overline{\ell}, \overline{\ell}'$ are the values of the mean anomalies corresponding to $\overline{u}, \overline{u}'$ (so that $\overline{\ell}' = 0$). We have used the following notations

$$
\mathcal{F} = \frac{ae\sin\omega}{\beta} ; \qquad\qquad \mathcal{G} = \frac{a(1 + e\cos\omega)}{\beta}
$$

and

$$
\overline{x} = \frac{a\beta^2}{1 + e\cos\omega}; \qquad \overline{x}' = a'; \qquad \overline{y} = \overline{z} = \overline{y}' = \overline{z}' = 0 .
$$

Definition 4. We call *Wetherill function* the approximated distance function d, whose square is defined by

$$
\mathrm{d}^2(\ell, \ell') = (r_1 - r_1')^2 + (r_2 - r_2')^2 + (r_3 - r_3')^2 =
$$
$$
= a'^2 k'^2 + \frac{a^2(1 + 2e\cos\omega + e^2)}{\beta^2}k^2 - 2kk'[\mathcal{G}a'\cos I] - 2d_{nod}^+\mathcal{F}k + (d_{nod}^+)^2
$$

with $k = \ell - \overline{\ell}, k' = \ell'$.

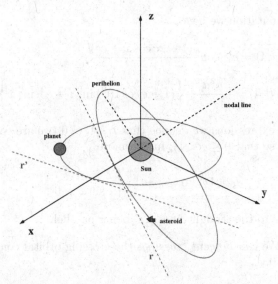

Fig. 2. The straight lines r, r′ represent Wetherill's approximation at the ascending node for the two osculating orbits of the asteroid and the planet.

Note that d^2 is a quadratic form in the variables k, k': it is homogeneous when there is a crossing at the ascending node. We can write it more concisely as

$$d^2(\ell, \ell') = d^2(\kappa) = \kappa^t \mathcal{A} \kappa + B^t \kappa + (d_{nod}^+)^2$$

where

$$\kappa = (k', k) ; \qquad B = 2(B_1, B_2) ; \qquad \mathcal{A} = \begin{bmatrix} A_{11} & A_{12} \\ A_{21} & A_{22} \end{bmatrix} ;$$

with components

$$\begin{cases} B_1 = 0 \\ B_2 = -d_{nod}^+ \mathcal{F} \end{cases} \qquad \begin{cases} A_{11} = a'^2 \\ A_{12} = A_{21} = -\mathcal{G}a' \cos I \\ A_{22} = [\mathcal{F}^2 + \mathcal{G}^2] . \end{cases}$$

For later use we define

$$\mathfrak{d}^2(u, u') = d^2(\ell(u), \ell'(u')) .$$

The geometry of Wetherill's straight lines is strictly related to the degeneracy of the matrix $\mathcal{A}$, in fact we have

Lemma 1. *The matrix $\mathcal{A}$ is always positive definite if $I > 0$. If $I = 0$ we have degeneracy of $\mathcal{A}$ if and only if the straight lines r, r′ are parallel: in this case $\mathcal{A}$ is positive semi-definite.*

Proof. $\mathcal{A}$ is a symmetric 2×2 matrix and it is positive definite if and only if its principal invariants, the trace $tr(\mathcal{A})$ and the determinant $det(\mathcal{A})$, are positive.

By a direct computation we have

$$\begin{cases} tr(\mathcal{A}) = a'^2 + a^2 \dfrac{(1 + 2e\cos\omega + e^2)}{1 - e^2} \\[2mm] det(\mathcal{A}) = \dfrac{a^2 a'^2}{(1 - e^2)} \left\{ (1 + e\cos\omega)^2 \sin^2 I + e^2 \sin^2 \omega \right\} . \end{cases}$$

From the above expressions we deduce that $tr(\mathcal{A}) > 0$ (we are considering only bounded orbits, so that $0 \le e < 1$); furthermore

$$det(\mathcal{A}) = 0 \quad \Longleftrightarrow \quad \begin{cases} I = 0 \\ e\sin\omega = 0 \, , \end{cases}$$

that corresponds to the straight lines r, r' being parallel.

Definition 5. We call *tangent crossings* the crossing orbital configurations for which $det(\mathcal{A}) = 0$.

The assumption that the inclination I of the asteroid is different from zero during its whole time evolution implies that *no tangent crossings occur*.

3.5   Kantorovich's method

We shall describe Kantorovich's method of singularity extraction (see [2]) that allows to improve the stability of the numerical computation of the integrals when the integrand function $f_1(x)$ is unbounded in the neighborhood of one or more points.

Kantorovich's method consists in searching for a function $f_2(x)$ whose primitive has an analytic expression in terms of elementary functions and such that the difference $f_1(x) - f_2(x)$ is more regular than $f_1(x)$ (for example it is bounded or even continuous).

It is then convenient to split the computation as follows

$$\int f_1(x)\,dx = \int [f_1(x) - f_2(x)]\,dx + \int f_2(x)\,dx$$

so that the singularity has moved to the second term, that can be better handled.

This method can help us to study the regularity properties of the averaged perturbing function $\overline{R}$ defined in (5); we shall use the inverse of the Wetherill function $1/d$ to extract the principal part from the direct term of the perturbing function.

The function D is 2π-periodic in both variables ℓ, ℓ' and this property can be used to shift the integration domain

$$\mathbb{T}^2 = \{(\ell, \ell') : -\pi \le \ell \le \pi, -\pi \le \ell' \le \pi\}$$

in a suitable way, so that the crossing values $(\overline{\ell}, 0)$ will be always internal points of this domain.

We shall prove that in computing the derivatives of $\overline{R}$ with respect to the variables $E_\mathcal{D}$, for instance the G-derivative, we can use the decomposition

$$\frac{(2\pi)^2}{\mu k^2} \frac{\partial}{\partial G} \overline{R} = \int_{\mathbb{T}^2} \frac{\partial}{\partial G} \left[\frac{1}{D} - \frac{1}{d} \right] d\ell \, d\ell' + \frac{\partial}{\partial G} \int_{\mathbb{T}^2} \left[\frac{1}{d} \right] d\ell \, d\ell'; \qquad (14)$$

namely we shall prove the validity of the hypotheses of the theorem of differentiation under the integral sign to exchange the symbols of integral and derivative in front of the *remainder function* $1/D - 1/d$. The average of the remainder function is then differentiable as it is derivable with continuity with respect to all the variables $E_\mathcal{D}$. Therefore we shall need only to study the regularity properties of the last term of the sum in (14), which is easier to handle.

Note that we use Kantorovich's method of singularity extraction in a wider extent: the derivatives of the remainder function still have a polar singularity in $(\overline{\ell}, 0)$, but it is of order one, so that the integrals over ℓ, ℓ' of these derivatives are convergent.

3.6 Integration of $1/d$

We shall discuss the analytic method to integrate $1/d$ over the unshifted domain $\mathbb{T}^2 = \{(\ell, \ell') : -\pi \leq \ell \leq \pi, -\pi \leq \ell' \leq \pi\}$, assuming that $(\overline{\ell}, 0)$ is an internal point of this domain.

We move the ascending node crossing point $(\overline{\ell}, 0)$ to the origin of the reference system by the variable change

$$\tau_{\overline{\ell}, 0} : (\ell, \ell') \longrightarrow (k, k') \qquad (15)$$

and we set

$$\overline{\mathbb{T}}^2 = \tau_{\overline{\ell}, 0} \left[\mathbb{T}^2 \right] = \{ (\ell - \overline{\ell}, \ell') : (\ell, \ell') \in \mathbb{T}^2 \} \ .$$

Then we perform another variable change to eliminate the linear terms in the quadratic form $d^2(\kappa)$ defined by (14). The inverse of the transformation used for this purpose is

$$\Xi^{-1} : \psi \longrightarrow \kappa = \mathcal{T} \psi + S \qquad (16)$$

where $S = (S_1, S_2) \in \mathbb{R}^2$, $\psi = (y', y) \in \mathbb{R}^2$ are the new variables and $\mathcal{T}$ is a 2×2 real-valued invertible matrix.

Setting to zero the coefficients of the linear terms of the quadratic form in the new variables ψ we obtain the equations

$$2 \, \mathcal{A} S + B = 0 \qquad (17)$$

whose solutions are

$$S_1 = \frac{B_2 A_{12}}{det(\mathcal{A})}; \qquad \qquad S_2 = -\frac{B_2 A_{11}}{det(\mathcal{A})} \ .$$

We can choose the matrix $\mathcal{T}$ such that

$$\mathcal{T}^t \mathcal{A} \mathcal{T} = \mathcal{I}_2$$

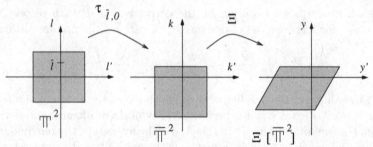

Fig. 3. Description of the transformations of the integration domain $\mathbb{T}^2$ with the two coordinate changes (15), (16) used to bring the squared Wetherill function $d^2(\ell, \ell')$ into the form $y^2 + y'^2 + (d_{min}^+)^2$ in the new variables (y', y). Note that $\overline{\ell}' = 0$ implies that $\overline{\mathbb{T}}^2$ is symmetric with respect to the k axis.

($\mathcal{I}_2$ is the 2×2 identity matrix) by setting

$$\mathcal{T} = \begin{pmatrix} 1/\tau & -\sigma/\tau\rho \\ 0 & 1/\rho \end{pmatrix}$$

with

$$\tau = \sqrt{A_{11}}; \qquad \rho = \sqrt{\frac{det(\mathcal{A})}{A_{11}}}; \qquad \sigma = A_{12}\sqrt{\frac{1}{A_{11}}} \;.$$

The coordinate change

$$\Xi : \kappa \longrightarrow \psi = \mathcal{R}\left[\kappa - S\right] \;, \tag{18}$$

where

$$\mathcal{R} = \mathcal{T}^{-1} = \begin{pmatrix} \tau & \sigma \\ 0 & \rho \end{pmatrix} \;,$$

brings $d^2(\kappa)$ into the form

$$d^2\left(\Xi^{-1}(\psi)\right) = y^2 + y'^2 + (d_{min}^+)^2$$

in the new variables ψ, with

$$d_{min}^+ = |d_{nod}^+|\left\{1 - \frac{a'^2\mathcal{F}^2}{det(\mathcal{A})}\right\}^{1/2} \;. \tag{19}$$

The domain $\overline{\mathbb{T}}^2$ is transformed into a parallelogram with two sides parallel to the y' axis (see Fig. 3).

Remark 4. Note that d_{min}^+ is the minimal distance between the straight lines r and r'.

Using the variable changes (15), (18) and the transformation to polar coordinates, whose inverse is

$$\Pi^{-1} : \begin{pmatrix} r \\ \theta \end{pmatrix} \longrightarrow \begin{pmatrix} y' \\ y \end{pmatrix} = \begin{pmatrix} r\cos\theta \\ r\sin\theta \end{pmatrix}$$

we obtain

$$\int_{\mathbb{T}^2} \frac{1}{d}\, d\ell\, d\ell' = \frac{1}{\sqrt{\det(\mathcal{A})}} \int_{\Xi[\overline{\mathbb{T}}^2]} \frac{1}{\sqrt{y^2 + y'^2 + (d_{min}^+)^2}}\, dy\, dy' =$$

$$(20)$$

$$= \frac{1}{\sqrt{\det(\mathcal{A})}} \int_{\mathfrak{I}} \frac{r}{\sqrt{r^2 + (d_{min}^+)^2}}\, dr\, d\theta$$

where $\Xi^{-1}\left[\Pi^{-1}(\mathfrak{I})\right] = \overline{\mathbb{T}}^2$.

Let us describe the domain $\mathfrak{I}$ in details. We define the straight lines that bound the integration domain $\Xi\left[\overline{\mathbb{T}}^2\right]$ as

$$\mathrm{r}_1 = \{(y,y') : y' = \frac{\sigma}{\rho}y + \tau(\pi - S_1)\}; \qquad \mathrm{r}_2 = \{(y,y') : y = \rho(\pi - \overline{\ell} - S_2)\};$$

$$\mathrm{r}_3 = \{(y,y') : y' = \frac{\sigma}{\rho}y - \tau(\pi + S_1)\}; \qquad \mathrm{r}_4 = \{(y,y') : y = -\rho(\pi + \overline{\ell} + S_2)\}\,.$$

The intersections of these lines with the y axis are

$$y_1 = -\rho(\pi + \overline{\ell} + S_2); \qquad y_2 = \rho(\pi - \overline{\ell} - S_2);$$

while the intersections with the y' axis are

$$y_1' = \lambda_3(y_1) = -\sigma(\pi + \overline{\ell} + S_2) - \tau(\pi + S_1)$$
$$y_2' = \lambda_3(0) = -\tau(\pi + S_1)$$
$$y_3' = \lambda_3(y_2) = \sigma(\pi - \overline{\ell} - S_2) - \tau(\pi + S_1)$$
$$y_4' = \lambda_1(y_1) = -\sigma(\pi + \overline{\ell} + S_2) + \tau(\pi - S_1)$$
$$y_5' = \lambda_1(0) = \tau(\pi - S_1)$$
$$y_6' = \lambda_1(y_2) = \sigma(\pi - \overline{\ell} - S_2) + \tau(\pi - S_1)$$

where $\lambda_1(y) = (\sigma/\rho)y + \tau(\pi - S_1)$ and $\lambda_3(y) = (\sigma/\rho)y - \tau(\pi + S_1)$.

We can then decompose the domain $\mathfrak{I}$ into four parts (see Fig. 4)

$$\mathfrak{I} = \bigcup_{j=1}^{4} \{(r,\theta) \in \mathbb{R}^2 \ : \ \theta_j \leq \theta \leq \theta_{j+1} \text{ and } 0 \leq r \leq r_j(\theta)\}$$

where $r_j(\theta)$, with $j = 1\ldots4$, represent the lines r_j delimiting $\Xi[\overline{\mathbb{T}}^2]$ in polar coordinates:

$$r_1(\theta) = \frac{\rho\tau(\pi - S_1)}{\rho\cos\theta - \sigma\sin\theta}; \qquad r_2(\theta) = \frac{\rho(\pi - \overline{\ell} - S_2)}{\sin\theta};$$

$$r_3(\theta) = \frac{-\rho\tau(\pi + S_1)}{\rho\cos\theta - \sigma\sin\theta}; \qquad r_4(\theta) = \frac{-\rho(\pi + \overline{\ell} + S_2)}{\sin\theta};$$

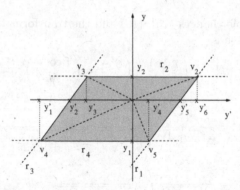

Fig. 4. We show the decomposition of the integration domain used to compute the last integral in (20) in polar coordinates

while $\theta_1 = \theta_5 - 2\pi$ and θ_l, with $l = 2 \ldots 5$, are the counter-clockwise angles between the y' axis and the vertexes v_l seen from the origin of the axes (see Fig. 4):

$$0 < \theta_2 < \theta_3 < \pi < \theta_4 < \theta_5 < 2\pi \, ;$$

$$\tan \theta_2 = \frac{\rho(\pi - \overline{\ell} - S_2)}{\sigma(\pi - \overline{\ell} - S_2) + \tau(\pi - S_1)} ; \quad \tan \theta_3 = \frac{\rho(\pi - \overline{\ell} - S_2)}{\sigma(\pi - \overline{\ell} - S_2) - \tau(\pi + S_1)} ;$$

$$\tan \theta_4 = \frac{\rho(\pi + \overline{\ell} + S_2)}{\sigma(\pi + \overline{\ell} + S_2) + \tau(\pi + S_1)} ; \quad \tan \theta_5 = \frac{\rho(\pi + \overline{\ell} + S_2)}{\sigma(\pi + \overline{\ell} + S_2) - \tau(\pi - S_1)} .$$

Using the previous decomposition for $\mathfrak{T}$ and integrating in the r variable the last expression in (20) we obtain

$$\int_{\mathbb{T}^2} \frac{1}{d} \, d\ell \, d\ell' = \frac{1}{\sqrt{det(\mathcal{A})}} \cdot \left\{ \sum_{j=1}^{4} \int_{\theta_j}^{\theta_{j+1}} \sqrt{\left(d_{min}^+\right)^2 + r_j^2(\theta)} \, d\theta - 2\pi d_{min}^+ \right\} . \quad (21)$$

Note that the integrals in (21) are elliptic and the integrand functions are bounded so that these integrals are differentiable functions of the orbital elements. We shall see that the loss of regularity of the averaged perturbing function is due only to the term d_{min}^+.

3.7 Boundedness of the remainder function

When there is a crossing at the ascending node, then from the equations of the orbits (8) and from Kepler's equations (10) we deduce that Taylor's development of $\mathcal{D}^2(\kappa) = D^2(\ell, \ell')$ in a neighborhood of $\kappa = (0, 0)$ is given by

$$\mathcal{D}^2(\kappa) = d^2(\kappa) + O(|\kappa|^3) \quad (22)$$

where $O(|\kappa|^3)$ is an infinitesimal of the same order as $|\kappa|^3$ for $|\kappa| \to 0$. We prove the following

Lemma 2. *If there is an ascending node crossing between the orbits, there exist a neighborhood $\mathcal{U}_0$ of $\kappa = (0,0)$ and two positive constants B_1, B_2 such that*

$$B_1 d^2(\kappa) \le \mathcal{D}^2(\kappa) \le B_2 d^2(\kappa) \qquad \forall \kappa \in \mathcal{U}_0.$$

Proof. First we notice that for $d_{nod}^+ = 0$ we have $d^2(\kappa) = \kappa^t \mathcal{A} \kappa$, where $\mathcal{A}$ is positive definite, hence there exist two positive constants C_1, C_2 such that

$$C_1 |\kappa|^2 \le \kappa^t \mathcal{A} \kappa \le C_2 |\kappa|^2 \qquad \forall \kappa \in \mathbb{R}^2. \tag{23}$$

Using the relations (22) and (23) we obtain

$$\lim_{|\kappa| \to 0} \frac{\mathcal{D}^2(\kappa)}{d^2(\kappa)} = 1,$$

that implies the existence of the neighborhood $\mathcal{U}_0$ and of the constants B_1, B_2 as in the statement of the lemma.

We prove the following result:

Proposition 1. *The remainder function $1/\mathrm{D} - 1/\mathrm{d}$ is bounded even if there is an ascending node crossing.*

Proof. If there are no crossings between the orbits the remainder function is trivially bounded, in fact $\mathrm{D}(\ell, \ell') > 0$ for each $(\ell, \ell') \in \mathbb{T}^2$ and the minimum value of $\mathrm{d}(\ell, \ell')$ is d_{min}^+ that, for $I \ne 0$, can be zero only if $d_{nod}^+ = 0$ (see equation (19)).

If there is a crossing at the ascending node we have to investigate the local behavior of the remainder function in a neighborhood of $(\ell, \ell') = (\bar{\ell}, 0)$, where both D and d can vanish. The boundedness of the remainder function can be shown using the previous lemma: we know that there exists a neighborhood $\mathcal{U}_0$ and a positive constant B_1 such that the relation

$$\mathcal{D}(\kappa) \ge \sqrt{B_1} d(\kappa)$$

holds for each $\kappa \in \mathcal{U}_0$. It follows that in this neighborhood the remainder function can be bounded in the following way:

$$\left| \frac{1}{\mathcal{D}(\kappa)} - \frac{1}{d(\kappa)} \right| = \frac{|d^2(\kappa) - \mathcal{D}^2(\kappa)|}{d(\kappa)\mathcal{D}(\kappa)[d(\kappa) + \mathcal{D}(\kappa)]} \le$$

$$\le \frac{1}{\sqrt{B_1}[1 + \sqrt{B_1}]} \cdot \frac{|d^2(\kappa) - \mathcal{D}^2(\kappa)|}{|\kappa|^3} \cdot \frac{|\kappa|^3}{d^3(\kappa)}.$$

We observe that $|d^2(\kappa) - \mathcal{D}^2(\kappa)| = O(|\kappa|^3)$ and that by (23) there is a positive constant C_1 such that $d^2(\kappa) \ge C_1 |\kappa|^2$, so that there exists a constant $L > 0$ such that

$$\left| \frac{1}{\mathcal{D}(\kappa)} - \frac{1}{d(\kappa)} \right| \le L \qquad \forall \kappa \in \mathcal{U}_0 .$$

Remark 5. Although the remainder function $1/D - 1/d$ is bounded, it is not continuous in $(\ell, \ell') = (\bar{\ell}, 0)$ when there is a crossing at the ascending node.

3.8   The derivatives of the averaged perturbing function $\overline{R}$

Kantorovich's method is used to describe the singularities of the derivatives of the averaged perturbing function with respect to Delaunay's variables appearing in equations (12).

Note that by the *chain rule* we can write

$$\frac{\partial \overline{R}}{\partial E_{\mathcal{D}}} = \frac{\partial \overline{R}}{\partial E_{\mathcal{K}}} \frac{\partial E_{\mathcal{K}}}{\partial E_{\mathcal{D}}}$$

where $E_{\mathcal{K}} = \{e, I, \omega, \Omega\}$ is a subset of the Keplerian elements of the asteroid and

$$\frac{\partial E_{\mathcal{K}}}{\partial E_{\mathcal{D}}} = \begin{bmatrix} \mathcal{M} & \mathcal{O} \\ \mathcal{O} & I_2 \end{bmatrix}$$

in which I_2 and $\mathcal{O}$ are the 2×2 identity and zero matrixes, and

$$\mathcal{M} = -\frac{1}{k\sqrt{a}} \begin{bmatrix} \beta/e & 0 \\ -\cot I/\beta & 1/(\beta \sin I) \end{bmatrix} .$$

Hence we can do the computation using the derivatives of $\overline{R}$ with respect to the Keplerian elements e, I, ω ($\overline{R}$ does not depend on Ω).

We shall not need to perform the splitting of Kantorovich's method to compute the derivative with respect to the inclination I; in fact the derivative of $1/\mathfrak{D}$ with respect to I can be bounded by a function with a first order polar singularity in $\overline{u}, \overline{u}'$, so it is Lebesgue integrable over $\mathbb{T}^2$.

In the following we shall first prove that the derivatives of the remainder function $1/D - 1/d$ are always Lebesgue integrable over $\mathbb{T}^2$, even if the two orbits intersect each other, so that the average of the remainder function is differentiable: indeed its derivatives can be computed by exchanging the position of the integral and differential operators as in (14). Then we shall see that if there is an ascending node crossing, then a discontinuous term appears in the derivatives of the average of $1/d$ and this is responsible of the discontinuity of the derivatives of $\overline{R}$. These derivatives admit two limit values at crossings (coming from the regions defined by $d_{nod}^+ > 0$ and $d_{nod}^+ < 0$).

As the properties we intend to prove are invariant by coordinate changes, we shall show them using the coordinates (u, u') instead of (ℓ, ℓ').

The derivatives of the remainder function $1/\mathfrak{D} - 1/\mathfrak{d}$.

Let us set $\upsilon = (u, u')$ and $\nu = (v, v') = (u - \overline{u}, u' - \overline{u}')$. We apply Taylor's formula with the integral remainder to the vector functions $P(u), P'(u')$:

$$\begin{cases} P(u) = P(\overline{u}) + P_u(\overline{u})\, v + \displaystyle\int_{\overline{u}}^{u} (u - s) P_{ss}(s)\, ds \\[2mm] P'(u') = P'(\overline{u}') + P'_{u'}(\overline{u}')\, v' + \displaystyle\int_{\overline{u}'}^{u'} (u' - t) P'_{tt}(t)\, dt . \end{cases}$$

The functions defining the straight lines $r(u) = \mathrm{r}(\ell(u))$ and $r'(u') = \mathrm{r}'(\ell'(u'))$ have the same Taylor's development, up to the first order in $|\nu| = \sqrt{v^2 + v'^2}$, as $P(u)$ and $P'(u')$ respectively, so that we can write

$$
\begin{cases}
r(u) = P(\overline{u}) + P_u(\overline{u})\, v + \displaystyle\int_{\overline{u}}^{u} (u - s) r_{ss}(s)\, ds \\[4mm]
r'(u') = P'(\overline{u}') + P'_{u'}(\overline{u}')\, v' + \displaystyle\int_{\overline{u}'}^{u'} (u' - t) r'_{tt}(t)\, dt \ .
\end{cases}
$$

We prove the following

Theorem 1. *If there is an ascending node crossing at $(u, u') = (\overline{u}, \overline{u}')$, the derivatives of the remainder function $1/\mathfrak{D} - 1/\eth$ with respect to e, ω can be bounded by functions having a first order polar singularity in $\overline{u}, \overline{u}'$, so they are Lebesgue integrable over $\mathbb{T}^2$.*

Proof. We shall consider only the derivatives with respect to e: the proof for the other derivatives is similar. First we note that

$$
\frac{\partial}{\partial e}\left[\frac{1}{\mathfrak{D}(v)}\right] = -\frac{1}{2\mathfrak{D}^3(v)} \frac{\partial}{\partial e}\left[\mathfrak{D}^2(v)\right] \ ; \qquad \frac{\partial}{\partial e}\left[\frac{1}{\eth(v)}\right] = -\frac{1}{2\eth^3(v)} \frac{\partial}{\partial e}\left[\eth^2(v)\right] \ .
$$

Let us write $\langle\, ,\, \rangle$ for the Euclidean scalar product. We have

$$
\frac{\partial}{\partial e}\left[\mathfrak{D}^2(v)\right] = \mathfrak{D}^2_{e,0}(v) + \mathfrak{D}^2_{e,1}(v) + \mathfrak{D}^2_{e,2}(v) \tag{24}
$$

where

$$
\mathfrak{D}^2_{e,0}(v) = 2\left\langle \frac{\partial}{\partial e}\left[P(u) - P'(u')\right], P(\overline{u}) - P'(\overline{u}')\right\rangle
$$

$$
\mathfrak{D}^2_{e,1}(v) = 2\left\langle \frac{\partial}{\partial e}\left[P(u) - P'(u')\right], P_u(\overline{u})\, v - P'_{u'}(\overline{u}')\, v'\right\rangle
$$

$$
\mathfrak{D}^2_{e,2}(v) = 2\left\langle \frac{\partial}{\partial e}\left[P(u) - P'(u')\right], \int_{\overline{u}}^{u} (u - s) P_{ss}(s)\, ds - \int_{\overline{u}'}^{u'} (u' - t) P'_{tt}(t)\, dt\right\rangle
$$

and

$$
\frac{\partial}{\partial e}\left[\eth^2(v)\right] = \eth^2_{e,0}(v) + \eth^2_{e,1}(v) + \eth^2_{e,2}(v) \tag{25}
$$

where

$$
\eth^2_{e,0(v)} = 2\left\langle \frac{\partial}{\partial e}\left[r(u) - r'(u')\right], P(\overline{u}) - P'(\overline{u}')\right\rangle
$$

$$
\eth^2_{e,1}(v) = 2\left\langle \frac{\partial}{\partial e}\left[r(u) - r'(u')\right], P_u(\overline{u})\, v - P'_{u'}(\overline{u}')\, v'\right\rangle
$$

$$
\eth^2_{e,2}(v) = 2\left\langle \frac{\partial}{\partial e}\left[r(u) - r'(u')\right], \int_{\overline{u}}^{u} (u - s) r_{ss}(s)\, ds - \int_{\overline{u}'}^{u'} (u' - t) r'_{tt}(t)\, dt\right\rangle \ .
$$

If we set the *crossing conditions* $P(\overline{u}) = P'(\overline{u}')$ we obtain

$$\mathfrak{D}^2_{e,0}(v) = \mathfrak{d}^2_{e,0}(v) = 0$$

and, in particular, the constant terms in Taylor's developments of $\partial \mathfrak{D}^2/\partial e$ and $\partial \mathfrak{d}^2/\partial e$ vanish.

The terms defined by $\mathfrak{D}^2_{e,2}$ and $\mathfrak{d}^2_{e,2}$ are at least infinitesimal of the second order with respect to $|v|$ as $v \to (\overline{u}, \overline{u}')$, so that the first order terms in $|v|$ at crossing can be given only by $\mathfrak{D}^2_{e,1}$ and $\mathfrak{d}^2_{e,1}$.

Using the theorems on the integrals depending on a parameter we obtain

$$\frac{\partial}{\partial e}\left[\int_{\overline{u}}^u (u-s)P_{ss}(s)\,ds - \int_{\overline{u}'}^{u'}(u'-t)P'_{tt}(t)\,dt\right] =$$

$$\int_{\overline{u}}^u (u-s)\frac{\partial P_{ss}}{\partial e}(s)\,ds - \frac{\partial \overline{u}}{\partial e}P_{uu}(\overline{u})\,v - \int_{\overline{u}'}^{u'}(u'-t)\frac{\partial P'_{tt}}{\partial e}(t)\,dt + \frac{\partial \overline{u}'}{\partial e}P'_{u'u'}(\overline{u}')\,v'$$

$$\frac{\partial}{\partial e}\left[\int_{\overline{u}}^u (u-s)r_{ss}(s)\,ds - \int_{\overline{u}'}^{u'}(u'-t)r'_{tt}(t)\,dt\right] =$$

$$\int_{\overline{u}}^u (u-s)\frac{\partial r_{ss}}{\partial e}(s)\,ds - \frac{\partial \overline{u}}{\partial e}r_{uu}(\overline{u})\,v - \int_{\overline{u}'}^{u'}(u'-t)\frac{\partial r'_{tt}}{\partial e}(t)\,dt + \frac{\partial \overline{u}'}{\partial e}r'_{u'u'}(\overline{u}')\,v'$$

so that these two expressions are at least infinitesimal of the first order with respect to $|v|$. As these terms are multiplied by first order terms in the expressions of $\mathfrak{D}^2_{e,1}$ and $\mathfrak{d}^2_{e,1}$, they give rise to at least second order terms.

We can conclude that the first order terms in the expressions (24) and (25) are equal and they are given by

$$2\left\langle \frac{\partial}{\partial e}[P(\overline{u}) - P'(\overline{u}')] - \left[\frac{\partial \overline{u}}{\partial e}P_u(\overline{u}) - \frac{\partial \overline{u}'}{\partial e}P'_{u'}(\overline{u}')\right], P_u(\overline{u})\,v - P'_{u'}(\overline{u}')\,v'\right\rangle ;$$

therefore the asymptotic developments of the e-derivatives of $\mathfrak{D}^2(v)$ and $\mathfrak{d}^2(v)$ in a neighborhood of $v = (\overline{u}, \overline{u}')$ are

$$\frac{\partial}{\partial e}\left[\mathfrak{D}^2(v)\right] = \alpha v + \beta v' + \mathfrak{r}_{\mathfrak{D}}(v); \qquad \frac{\partial}{\partial e}\left[\mathfrak{d}^2(v)\right] = \alpha v + \beta v' + \mathfrak{r}_{\mathfrak{d}}(v)$$

where α, β are independent on u, u' and $\mathfrak{r}_{\mathfrak{D}}(v), \mathfrak{r}_{\mathfrak{d}}(v)$ are infinitesimal of the second order with respect to $|v|$ as $v \to (\overline{u}, \overline{u}')$.

Using the decomposition

$$\left[\frac{1}{\mathfrak{D}^3} - \frac{1}{\mathfrak{d}^3}\right] = \left[\frac{1}{\mathfrak{D}} - \frac{1}{\mathfrak{d}}\right]\left[\frac{1}{\mathfrak{D}^2} + \frac{1}{\mathfrak{D}\mathfrak{d}} + \frac{1}{\mathfrak{d}^2}\right],$$

the boundedness of the remainder function $1/\mathfrak{D} - 1/\mathfrak{d}$ and lemma 2 (that also hold in the (u, u') coordinates), we conclude that there exist two constants $L_1, L_2 > 0$ such that

$$\left|\frac{\partial}{\partial e}\left[\frac{1}{\mathfrak{D}(v)}\right] - \frac{\partial}{\partial e}\left[\frac{1}{\mathfrak{d}(v)}\right]\right| = \frac{1}{2}\left|\left\{\left[\frac{1}{\mathfrak{D}^3(v)} - \frac{1}{\mathfrak{d}^3(v)}\right](\alpha v + \beta v') + \right.\right.$$

$$\left.\left. + \frac{1}{\mathfrak{D}^3(v)}\mathfrak{r}_{\mathfrak{D}}(v) - \frac{1}{\mathfrak{d}^3(v)}\mathfrak{r}_{\mathfrak{d}}(v)\right\}\right| \le L_1\frac{1}{|v|} + L_2$$

in a neighborhood of $\upsilon = (\overline{u}, \overline{u}')$ and the theorem is proven.

Singularities of the $\{e, \omega\}$-derivatives of the average of $1/d$.

As $det(\mathcal{A}) > 0$ and $(\overline{\ell}, 0)$ is in the interior part of $\mathbb{T}^2$, we have $d_{min}^2 + r_j^2(\theta) > 0$ for each $\theta \in [\theta_j, \theta_{j+1}]$ and for each $j = 1 \ldots 4$. Then we can use again the theorem of differentiation under the integral sign and compute, for instance, the derivative of the average of $1/d$ with respect to e as

$$\frac{\partial}{\partial e} \int_{\mathbb{T}^2} \frac{1}{d} \, d\ell \, d\ell' = \frac{\partial}{\partial e} \left[\frac{1}{\sqrt{det(\mathcal{A})}} \right] \cdot \left\{ \sum_{j=1}^{4} \int_{\theta_j}^{\theta_{j+1}} \sqrt{\left(d_{min}^+\right)^2 + r_j^2(\theta)} \, d\theta - 2\pi d_{min}^+ \right\} +$$

$$+ \left[\frac{1}{\sqrt{det(\mathcal{A})}} \right] \cdot \left\{ \frac{1}{2} \sum_{j=1}^{4} \int_{\theta_j}^{\theta_{j+1}} \frac{\frac{\partial}{\partial e}[(d_{min}^+)^2 + r_j^2(\theta)]}{\sqrt{\left(d_{min}^+\right)^2 + r_j^2(\theta)}} \, d\theta - 2\pi \frac{\partial}{\partial e} d_{min}^+ \right\} .$$

(26)

We have similar formulas for the derivatives with respect to ω, obtained simply by substitution of the partial derivative operators.

The discontinuities present in the terms

$$\frac{\partial}{\partial e} d_{min}^+ ; \qquad \frac{\partial}{\partial \omega} d_{min}^+ ;$$

are responsible of the discontinuities in the derivatives of the averaged perturbing function that produce a sort of *crests* in the surfaces representing this function (see Fig. 5) and cause the loss of regularity in its level lines, where the weak averaged solutions lie. The detailed analytical formulas for the discontinuities in the derivatives of $\overline{R}$ can be found in [10], [6].

4   Secular evolution theory

The generalized averaging principle has been used in [10] to define a method to compute the secular evolution of the NEAs in the framework of a Solar System with the planets on circular coplanar orbits. We shall review this method in the following of this section and we shall describe some features of the secular dynamics of NEAs.

First we note that the averaged Hamiltonian $\overline{H}$ *is invariant under the symmetries*

$$\omega \longrightarrow -\omega$$
$$\omega \longrightarrow \pi - \omega$$
$$\omega \longrightarrow \pi + \omega ;$$

this allows to draw the level lines of the averaged Hamiltonian, on which the solution curves are confined, simply knowing their shape in a subset of the reduced phase space (ω, e) of the form

$$\{(\omega, e) : k\,\pi/2 \le \omega \le (k+1)\,\pi/2 \,, \; 0 \le e \le e_{max}\}$$

with $k \in \mathbb{Z}$.

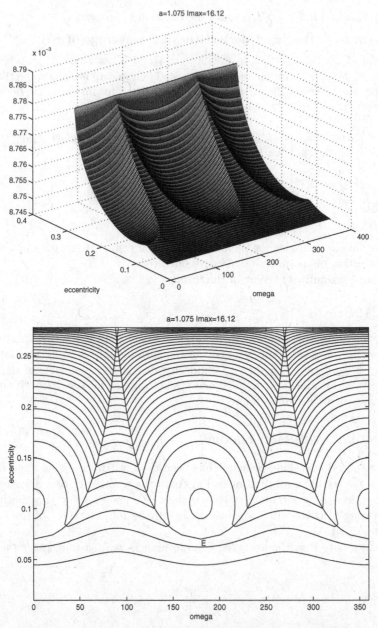

Fig. 5. We draw the graphic of the averaged perturbing function (top) and its level lines (bottom) in the plane (ω, e) for the Near Earth Asteroid 2000 CO_{101} (ω is in degrees in the figure). The loss of regularity at the node crossing lines with the Earth is particularly evident for this object.

4.1   The secular evolution algorithm

The numerical method used in [10] to solve Hamilton's equations (12) is an implicit Runge-Kutta-Gauss algorithm of order 6, which is also symplectic (see [23]).

Note that Kantorovich's method, used to study the regularity of the averaged perturbing function, gives us analytical formulas for the discontinuities of the derivatives at the right hand sides of (12).

The Runge-Kutta-Gauss methods use sub-steps which include neither the starting point nor the final point of the step being computed, this allows to avoid the computation of the values of the derivatives of $\overline{R}$ at the node crossing points, where they are defined in a twofold way. We resort to the following procedure (which had already been used in [18]): every time the asteroid orbit is close enough to a node crossing line, the standard iteration scheme known as *regula falsi* is used to set the second extreme point of the step exactly on that line; as the computation of the right hand sides of (12) is performed only at the intermediate points of the integration step, we avoid the computation at node crossings.

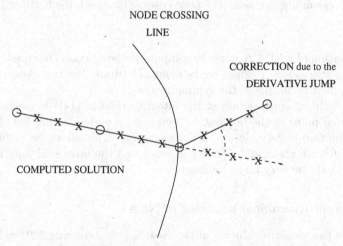

NODE CROSSING

LINE

CORRECTION due to the

DERIVATIVE JUMP

COMPUTED SOLUTION

Fig. 6. Graphical description of the algorithm employed in this numerical integration: it is an implicit Runge-Kutta-Gauss method, symplectic, of order 6. In an integration step, delimited in the figure by two consecutive small circles, we compute the derivatives of $\overline{R}$ only at the intermediate points marked with crosses.

When the node crossing point is reached within the required precision, then a correction given by the explicit formulas for the discontinuities of the derivatives of $\overline{R}$ is applied before restarting the integration (see Fig. 6).

An additional regula falsi is used to compute the value of the solutions of the averaged equations exactly at the symmetry lines in the plane (ω, e) (that is at the lines of the form $\omega = k\pi/2$ with $k \in \mathbb{Z}$); the computation of the

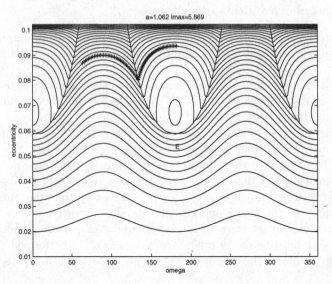

Fig. 7. Secular evolution figure for 2001 QJ_{142} on the background of the level lines of the averaged perturbing function. The node crossing lines with the Earth (E) are also drawn.

secular evolution of a NEA requires to compute the solution of the equations (12) between two successive crossings of the symmetry lines. The complete evolution is then obtained by means of the symmetries of $\overline{R}$.

Kantorovich's decomposition of the integrals (like in (14)) is used not only when the final point of the integration step is on a node crossing line, but also when the solution is very close to a node crossing. This allows to stabilize the computation when the nodal distance is small and the integrand functions can be bounded only by very large constants.

4.2   Different dynamical behavior of NEAs

We describe the secular evolution of the Near Earth Asteroids 2001 QJ_{142} and 1999 AN_{10}: these celestial objects show two very different kinds of dynamical behavior.

In Fig. 7 we can see that the perihelion argument ω of 2001 QJ_{142} circulates; the loss of regularity of the solution curve, corresponding to a crossing at the ascending node with the Earth, is particularly enhanced in this figure.

In Fig. 8 we have ω-libration for 1999 AN_{10}. Note that it starts its secular evolution with a double crossing with the orbit of the Earth, that gives it two possibilities to approach the Earth for each revolution and makes this object particularly dangerous. This asteroid has been intensively studied as a possible Earth impactor for the years 2039 (see [16]) and 2044 (see [17]).

In addition to this kind of libration, symmetric with respect to the lines $\omega = k\pi/2$ ($k \in \mathbb{Z}$), it is possible to have a sort of asymmetric librations, as it is shown by some of the level lines of Fig. 9 in which the evolution of the asteroid (2100) Ra Shalom is shown. It is possible to choose initial values for e and ω, defining a fictitious object with the same value of the integral a as (2100) Ra Shalom, such that an asteroid starting with those values is constrained into a very narrow asymmetric libration.

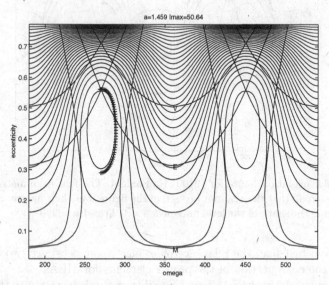

Fig. 8. Secular evolution figure for 1999 AN$_{10}$. The node crossing lines with the Earth (E) and Venus (V) are drawn.

We remark that asymmetric librations have a very low probability to occur: all the known NEAs examined so far do not show this kind of behavior.

5   Proper elements for NEAs

In the study of orbit dynamics a very important role is played by the integrals of the motion, that is by quantities that are constant during the time evolution of a dynamical system.

When the dynamics is non integrable, as it is the case for the N-body problem ($N \geq 3$), it is also useful to compute quantities that are nearly constant during the motion. We give the following definition:

Definition 6. The *proper elements* are quasi–integrals of the motion, that is quantities that change very slowly with time and can be considered approximatively constant over time spans not too long.

The first to employ the concept of proper elements was Hirayama [12]: he defined a linear theory to identify asteroid families in the main belt. The identification of families together with the understanding of the dynamical structure

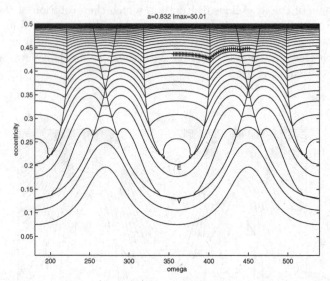

Fig. 9. Secular evolution figure for (2100) Ra Shalom. The four asymmetric libration regions are so small that they cannot be seen in the figure, but their presence can be detected by seeing the shape of the level lines near $e = 0.25$ and $\omega = 240°, 300°, 420°, 480°$.

of the asteroid belt (e.g. the relevance of secular resonances) are two important reasons for the computation of the proper elements for MBAs.

There are presently three different possible methods to compute these quantities:

1. An analytical theory by Milani and Knežević [19],[20],[21] based on series expansion in eccentricity and inclination, particularly suitable for orbits with low eccentricity and low inclination ($< 17°$). The proper elements computed in [20],[21] have been proven to be stable over time scales of the order of 10^7 years.
2. A semianalytic theory by Lemaître and Morbidelli [15], which is more appropriate for the orbits with either large eccentricities or large inclinations. This method is based on the classical averaging method in a revisited version by Henrard [11]. A similar method had already been used by Williams [26],[27],[28] to obtain a set of proper elements that led to the understanding of the secular resonances resulting from two secular frequencies being equal.
3. A synthetic theory by Knežević and Milani [14] based on the computation of the asteroid orbits by pure numerical integration; the short periodic perturbations are then removed by a filtering process performed during the integration. This recent theory allows to obtain a high accuracy of the elements; on the other hand it requires CPU times longer than the two preceding methods.

Almost all the NEAs are planet crossing, and the singularities coming from the possibility of collisions result in the strong divergence of the series used in the analytical theory and in the divergence of the integrals of the classical averaging principle used in the semianalytic theory. Furthermore the strong chaoticity of crossing orbits and the very short integration steps to be chosen in this case make the synthetic theory inapplicable on large scale: in fact we have to use several values of the initial phase on the orbit for each NEA and in this case it would require very long computational times.

On the other hand the reasons to compute proper elements are very different in the case of NEAs and the required stability times are only of the order of 10 000 to 100 000 years. Over longer time spans the dynamics is dominated by large changes in the orbital elements, including the semimajor axis, resulting from the close approaches with the planets and from the effects of secular resonances.

We wish to compute proper elements for NEAs mainly for the following reasons:

1. to detect the possibility of collision of Earth-crossing objects and to compute its probability;
2. to identify the objects whose long term evolution is controlled by one or more of the main secular resonances;
3. to identify meteor streams (sets of very small objects, that can be observed only when they are crossing the orbit of the Earth) and to give a criterion to be used in the identification of their parent bodies.

We give the definition of proper elements for Near Earth Asteroids and we explain how to compute them using the generalized averaging principle explained in section 3.

Given the osculating orbit of a NEA, represented by its Keplerian elements $(a, e, I, \omega, \Omega)$ at a given time t, the following quantities are constant during the averaged motion in the framework of the circular coplanar case:

$$a, e_{min}, e_{max}, I_{min}, I_{max} \; ;$$

they are respectively the semimajor axis and the minimum and maximum value of the averaged eccentricity and the averaged inclination.

We can consider as set of proper elements either

$$\{a, e_{min}, I_{max}\} \qquad \text{or} \qquad \{a, e_{max}, I_{min}\} \; .$$

Remark 6. If we consider an ω-librating orbit we can also define

$$\omega_{min}, \omega_{max},$$

that are the minimum and maximum value of the averaged perihelion argument. These quantities are also constant during the averaged motion and can give additional informations to understand the dynamics of the objects that we are studying.

We can also use a set of proper elements in which the extreme values of the eccentricity and inclination are substituted by the secular frequencies of the longitude of perihelion g and the longitude of the node s in case of ω-circulation. If ω is librating we can use the libration frequency l_f in place of g.

The computation of these proper frequencies requires some additional comments:

Proposition 2. *If ω is circulating, then let t_0 and t_1 be the times of passage at 0 and $\pi/2$ and let $\Omega_{t_0}, \Omega_{t_1}$ be the corresponding values of the longitude of the node. We have the following formulas to compute the secular frequencies of the argument of perihelion $g - s$ and of the longitude of the node s:*

$$g - s = \frac{2\pi}{4(t_1 - t_0)}; \qquad s = \frac{\Omega_{t_1} - \Omega_{t_0}}{t_1 - t_0} .$$

If ω is symmetrically librating, then let τ_0 and τ_1 be two consecutive times of passage at the same integer multiple of $\pi/2$ and let $\Omega_{\tau_0}, \Omega_{\tau_1}$ be the corresponding values of the longitude of the node. We can compute the frequency of the longitude of the node s and the libration frequency l_f by the following formulas:

$$s = \frac{\Omega_{\tau_1} - \Omega_{\tau_0}}{\tau_1 - \tau_0}; \qquad l_f = \pm\frac{2\pi}{2(\tau_1 - \tau_0)} ,$$

where the sign has to be chosen negative for clockwise libration.

Proof. The formula for $g - s$ is an immediate consequence of the fact that the period of circulation of ω must be four times the time interval required by an increase by $\pi/2$ of ω.

The proper frequency s has to be computed taking into account that the node has a secular precession, but also long periodic oscillations controlled by the argument 2ω.

The averaged Hamiltonian does not contain Ω because of the invariance with respect to rotation around the z axis; according to D'Alembert's rules (see [24]) it contains only the cosine of 2ω, thus the perturbative equation of motion for Ω is

$$\frac{d\Omega}{dt} = -\frac{\partial \overline{R}}{\partial Z}(2\omega) .$$

If ω is circulating it is possible to change (at least locally) variable and write the equation

$$\frac{d\Omega}{d\omega} = \frac{\partial \overline{R}/\partial Z}{\partial \overline{R}/\partial G} = F(\cos(2\omega))$$

with a right hand side containing only $\cos(2\omega)$. The solution for Ω as a function of ω is:

$$\Omega(\omega) = \alpha \, \omega + f(\sin(2\omega))$$

and the function $f(\sin(2\omega))$ is zero for $\omega = 0, \pi/2$, thus

$$\Omega_{t_1} - \Omega_{t_0} = \alpha \, \pi/2$$

identifies the secular part of the evolution of Ω, whose frequency can be computed by

$$s = \frac{\Omega_{t_1} - \Omega_{t_0}}{t_1 - t_0} .$$

The symmetric libration cases are different because ω returns to the same multiple of $\pi/2$ after a time interval $\tau_1 - \tau_0$, while in the same time span Ω has changed by $\Omega_{\tau_1} - \Omega_{\tau_0}$. Then $g - s$ is not a proper frequency (its secular part is by definition zero), and by arguments that are similar to the ones above, the libration frequency l_f and the secular frequency of the node s can be computed by

$$l_f = \pm\frac{2\pi}{2(\tau_1 - \tau_0)}; \qquad s = \frac{\Omega_{\tau_1} - \Omega_{\tau_0}}{\tau_1 - \tau_0} .$$

We agree that the negative sign for l_f is chosen for clockwise librations.

6   Reliability tests

A numerical test on the reliability of the weak averaged solutions and of the proper elements for NEAs obtained with the generalized averaging principle can be found in [7]. A sample of orbits has been taken into account and the weak averaged solutions for the objects of this sample have been compared with their corresponding quantities obtained by pure numerical integrations using initial conditions that correspond to circular coplanar orbits for the planets.

The results of this comparison are satisfactory: the cases in which the secular evolution theory fails are generally the ones for which it is not valid *a priori*, that is low order mean motion resonances and close approaches with one or more planets, that could change the value of the semimajor axis, which is assumed to be constant in the averaging theory.

Table 1. Proper elements table for the asteroids (433) Eros and (1981) Midas: the angles are given in degrees. In case of ω-circulation we define $\omega_{min} = -\infty, \omega_{max} = +\infty$. AT means Averaged Theory, while NI means Numerical Integrations.

Prop. El.	(433) *Eros*		(1981) *Midas*	
	AT	NI	AT	NI
e_{min}	0.22285	0.22272	0.34872	0.34978
e_{max}	0.23307	0.23289	0.65075	0.65029
I_{min}	10.06653	10.06472	39.78069	39.79393
I_{max}	10.82960	10.83039	51.49418	51.43053
ω_{min}	$-\infty$	$-\infty$	248.61698	248.51999
ω_{max}	$+\infty$	$+\infty$	291.38302	291.56100

The agreement of the elements for the sample selected in [7] is of the order of 5×10^{-3} both for the proper eccentricity and the proper inclination (if considered in radians) over a time span of the order of 10^4 yrs.

Table 2. Proper frequencies table for (433) Eros and (1981) Midas: the units are arc-seconds per year. In case of ω-circulation we define $l_f = 0$ while in case of ω-libration we define $g = s$.

	(433) *Eros*		(1981) *Midas*	
Prop. Freq.	AT	NI	AT	NI
g	14.675	14.876	−25.625	−25.650
s	−21.101	−21.136	−25.625	−25.650
l_f	0	0	−45.672	−45.422

In Tables 1,2 we present the set of proper elements and proper frequencies of two objects of the sample: 433 (Eros) and (1981) Midas, that are respectively ω-circulating and ω-librating. The orbital elements used for these asteroids are in Table 3.

Table 3. Orbital elements used in the comparison for (433) Eros and for (1981) Midas.

Body	a (AU)	ω (°)	Ω (°)	e	I (°)
(433) Eros	1.45823	178.640	304.411	0.222863	10.829
(1981) Midas	1.77611	267.720	357.097	0.650113	39.831

In [7] there is also a comparison of the previous results with full numerical integrations starting with the actual eccentricity and inclination of the planets: we observe that the difference between the proper elements obtained in both ways for the selected sample is not dramatic (at least for the time scale of this integration), but the crossing times between the orbits are not reliable at all if we do not take into account the eccentricity and inclination of the planets and the computation of these times is useful for several applications such as the study of the possibility of collision. Thus the need of a more accurate averaging theory is evident.

7   Generalized averaging principle in the eccentric–inclined case

Recently we have proven that the generalized averaging theory, defined in Section 3, can be extended including the eccentricities and the inclinations of the planets. We give in this section only the main idea of this generalization; the reader interested can found a complete explanation of the theory in [5],[6].

Also in this framework we can study the case of only one perturbing planet and we can obtain the total perturbation by the sum of the contribution of each planet in the model. We can write the averaged perturbing function $\overline{R}$ as a function of a particular set of variables, called *mutual elements*, that are almost everywhere regular functions of Delaunay's variables and are defined by the mutual position of the osculating orbits of the asteroid and the planet. Then

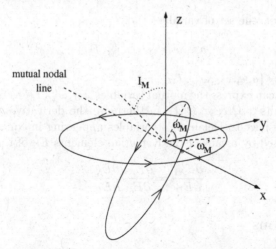

Fig. 10. The mutual reference frame: the ascending mutual node is marked with asterisks.

the equations of motion for the asteroid become

$$
\begin{cases}
\dot{\tilde{G}} = \dfrac{\partial \overline{R}}{\partial E_{\mathcal{M}}} \dfrac{\partial E_{\mathcal{M}}}{\partial g} \\[2ex]
\dot{\tilde{Z}} = \dfrac{\partial \overline{R}}{\partial E_{\mathcal{M}}} \dfrac{\partial E_{\mathcal{M}}}{\partial z}
\end{cases}
\qquad
\begin{cases}
\dot{\tilde{g}} = -\dfrac{\partial \overline{R}}{\partial E_{\mathcal{M}}} \dfrac{\partial E_{\mathcal{M}}}{\partial G} \\[2ex]
\dot{\tilde{z}} = -\dfrac{\partial \overline{R}}{\partial E_{\mathcal{M}}} \dfrac{\partial E_{\mathcal{M}}}{\partial Z}
\end{cases}
\tag{27}
$$

where $E_{\mathcal{M}}$ is a suitable subset of the mutual elements.

First note that the definitions of *mutual nodal line* and *mutual nodes* make sense even in the elliptic case; but we observe that we have generally a different mutual nodal line for each planet while in the circular coplanar case it is the same for all of them.

7.1   The mutual reference frame

We give a short description of the mutual elements. Let us consider two elliptic non-coplanar osculating orbits of an asteroid and a planet with a common focus: we give the following

Definition 7. We call *mutual reference frame* a system $Oxyz$ (see Fig. 10) such that the x axis is along the mutual nodal line and is directed towards the mutual ascending node; the y axis lies on the planet orbital plane, so that the orbit of the planet lies on the (x, y) plane. We shall use the further convention that the positive z axis is oriented as the angular momentum of the planet.

Let ω_M, ω'_M be the mutual pericenter arguments (the angles between the x axis and the pericenters) of the orbit of the asteroid and of the planet respectively, and let I_M be the mutual inclination between the two conics. We define as *mutual elements* the set of variables

$$\{a, e, a', e', \omega_M, \omega'_M, I_M\}$$

and we set $E_{\mathcal{M}} = \{a, e, \omega_M, \omega'_M, I_M\}$.

Note that we can express the mutual variables ω_M, ω'_M, I_M as functions of the Keplerian elements $\omega, \Omega, I, \omega', \Omega', I'$; furthermore the derivatives of the mutual elements with respect to Delaunay's variables appearing in equations (27) can be easily computed by means of the Keplerian elements $E_{\mathcal{K}}$ of the asteroid:

$$\frac{\partial E_{\mathcal{M}}}{\partial E_{\mathcal{D}}} = \frac{\partial E_{\mathcal{M}}}{\partial E_{\mathcal{K}}} \frac{\partial E_{\mathcal{K}}}{\partial E_{\mathcal{D}}} .$$

8   Conclusions

The generalized averaging principle and the related results reviewed in this paper have been applied to the search for parent bodies of meteor streams (using appropriate variables, like the ones in [25]) and to the computation of the secular evolution of the MOID (using algebraic methods as in [4]). We think that the extension of the theory including the eccentricity and inclination of the planets will be very useful to improve the accuracy of the results of such applications. There are additional possible applications of this theory, that we have still to investigate, to the computation of the collision probability between a NEA and the Earth.

Acknowledgements

The author wish to thank A. Celletti and A. Milani for their very useful suggestions in writing this paper.

References

1. ARNOLD, V.: 1997, *Mathematical Aspects of Classical and Celestial Mechanics*, Springer-Verlag, Berlin Heidelberg
2. DEMIDOVIC, B.P. AND MARON I.A.: 1966. *Foundations of Numerical Mathematics*, SNTL, Praha
3. FLEMING, W. H.: 1964. *Functions of Several Variables*, Addison-Wesley
4. GRONCHI, G. F.: 2002, *On the stationary points of the squared distance between two ellipses with a common focus*, SIAM Journal on Scientific Computing, to appear
5. GRONCHI, G. F.: 2002, *Generalized averaging principle and the secular evolution of planet crossing orbits*, Cel. Mech. Dyn. Ast., to appear
6. GRONCHI, G.F.: 2002. 'Theoretical and computational aspects of collision singularities in the N-body problem', *PHD Thesis, University of Pisa*, in preparation
7. GRONCHI, G. F. AND MICHEL, P.: 2001, *Secular Orbital Evolution, Proper Elements and Proper Frequencies for Near-Earth Asteroids: A Comparison between Semianalytic Theory and Numerical Integrations*, Icarus **152**, pp. 48-57

8. GRONCHI, G. F. AND MILANI, A.: 1998, *Averaging on Earth-crossing orbits*, Cel. Mech. Dyn. Ast. **71/2**, pp. 109-136

9. GRONCHI, G. F. AND MILANI, A.: 1999, *The stable Kozai state for asteroids and comets with arbitrary semimajor axis and inclination*, Astron. Astrophys. **341**, p. 928-935

10. GRONCHI, G. F., AND MILANI, A.: 2001, *Proper elements for Earth crossing asteroids*, Icarus **152**, pp. 58-69

11. HENRARD, J.: 1990, *A semi–numerical perturbation method for separable Hamiltonian systems*, Cel. Mech. Dyn. Ast. **49**, pp. 43-67

12. HIRAYAMA, K.: 1918, *Groups of asteroids probably of common origin*, Astron. J. **31** pp. 185-188

13. KOZAI, Y.: 1962, *Secular perturbation of asteroids with high inclination and eccentricity*, Astron. J. **67**, pp. 591-598

14. KNEZEVIĊ, Z. AND MILANI, A.: 2000, *Synthetic proper elements for outer main belt asteroids*, Cel. Mech. Dyn. Ast. **78 (1/4)** pp. 17-46

15. LEMAITRE, A. AND MORBIDELLI, A.: 1994, *Proper elements for high inclined asteroidal orbits*, Cel. Mech. Dyn. Ast. **60/1**, pp. 29-56

16. MILANI, A., CHESLEY, S. R. AND VALSECCHI, G. B.: 1999, *Close approaches of asteroid 1999 AN10: resonant and non-resonant returns*, Astron. Astrophys., **346**, pp. 65-68

17. MILANI, A., CHESLEY, S. R. AND VALSECCHI, G. B.: 2000, *Asteroid close encounters with Earth: risk assessment*, Plan. Sp. Sci., **48**, pp. 945-954

18. MILANI, A. AND GRONCHI, G. F.: 1999, 'Proper elements for Earth-crossers', in *Evolution and Source Regions of Asteroids and Comets*, J. Svoreň, E. M. Pittich, and H. Rickman eds., pp. 75-80

19. MILANI, A. AND KNEZEVIĊ, Z.: 1990, 'Secular perturbation theory and computation of asteroid proper elements', Cel. Mech. Dyn. Ast. **49**, pp. 347-411

20. MILANI, A. AND KNEZEVIĊ, Z.: 1992, *Asteroid Proper Elements and Secular Resonances*, Icarus **107**, pp. 219-254

21. MILANI, A. AND KNEZEVIĊ, Z.: 1994, *Asteroid Proper Elements and the Dynamical Structure of the Asteroid Main Belt*, Icarus **107**, pp. 219-254

22. MORBIDELLI, A. AND HENRARD, J.: 1991, *Secular resonances in the asteroid belt - Theoretical perturbation approach and the problem of their location* Cel. Mech. Dyn. Ast. **51/2**, pp. 131-167

23. SANZ-SERNA, J.M.: 1988, *Runge-Kutta schemes for Hamiltonian systems*, BIT **28**, pp. 877-883

24. SZEBEHELY, V.: 1967, *Theory of orbits*, Academic Press, New York and London

25. VALSECCHI, G. B., JOPEK, T. J. AND FROESCHLÈ, CL.: 1999, *Meteoroid streams identification: a new approach*, MNRAS **304**, pp. 743-750

26. WILLIAMS, J. G.: 1969, *Secular perturbations in the Solar System*, Ph.D. dissertation, University of California, Los Angeles

27. WILLIAMS, J. G.: 1979, *Proper elements and family membership of the asteroids*, In *Asteroids* (T. Gehrel, Ed.) pp. 1040-1063. Univ. Arizona Press, Tucson

28. WILLIAMS, J. G., AND FAULKNER, J.: 1981, *The position of secular resonances surfaces*, Icarus **46**, pp. 390-399

Subject Index

B_3 transformation 49

Amors 8
analytic continuation 96
angular momentum 82
Apollos 8
approach to equilibrium 109, 111, 112
asteroids 4
Atens 8
averaging principle 179, 181, 183, 187, 199, 207, 208, 210

b-plane 150
Balescu-Lenard equation 110
BBGKY hierarchy 105, 106, 110
Big Bang 3
bilinear relation 41
Birkhoff transformation 49
black holes 3
Boltzmann equation 110
Boltzmann–Grad limit 110

canonical transformation 30, 40, 63, 66–68, 70
capture 120, 122
center manifolds 96
central configurations 17, 86
chaos 103
close encounters 145
collision 25, 103
collision singularities 73–75, 79, 81
collisions 4, 63, 104
comets 4
correlation function 105
Coulomb potential 109, 111

D'Alembert 206
Delaunay's variables 180, 182, 196, 208, 210

distinguished limit 95
distribution function 105
double crossing 185, 187, 202
dynamics existence 102

Easton's method 10
Edgeworth-Kuiper 5
eigenvalues 91
elliptic integral 194
encounter 115, 117, 120
energy perturbations 146
entropy 108, 110–112
Euler's theorem on homogeneous functions 82
Eulerian central configuration 91
extended phase space 10, 31

fictitious time 10, 31, 36, 63, 64, 66, 67, 69, 84
finite range potential 102

generalized eccentric anomaly 63, 65, 68
generating function 27
geometric constraints 89
gravitational interactions 145

H-theorem 108, 110–112
Hamilton-Jacobi transformation 63, 67–69
Hamiltonian 82
hard spheres 102, 111
harmonic oscillator 32
heteroclinic 125, 127, 132
Hill's coordinates 117
Hill's equations 116, 119, 120
Hill's problem 19
homoclinic 125, 127, 132, 133
homographic solutions 18
homothetic solutions 18, 86

impact parameter   115, 120–122, 134
inclined billiard   18
inner solution   96
integrable system   103
invariant manifolds   96

Jacobi constant   29
Jacobi integral   29
Jacobi matrix   91

K-S transformation   63, 65, 66, 69
Kantorovich   190, 196, 201
Kepler motion   87
kinetic theory   105
Kozai   184, 185
Kustaanheimo–Stiefel   25, 36

Lagrange multipliers   89
Lagrange's formula   83
Lagrangean configuration   90
Levi–Civita   25, 29
Levi–Civita transformation   10, 12
Levy-Civita   120, 121
Liouville equation   106
Lorentz gas   101
Lyapunov exponent   103, 104, 108

Maxwell potential   111
mean motion resonances   181, 183, 207
mean-field limit   107
Meteor Crater   5
meteorites   5
meteoroid streams   145
meteoroids   5
meteors   5
Minimum Orbital Intersection Distance
  (MOID)   155
mixing   109
MOID   8, 210
molecular chaos   110
moment of inertia   83
multiple collisions   74–78
mutual elements   208, 209
mutual reference frame   209

Near–Earth Asteroids   8
Near–Earth Comets   8
Near–Earth Objects   5
Near-Earth Asteroids   145
NEODyS   179

Newton potential   109
nodal distance   145, 184, 202
non-collision singularities   20, 72–74, 77,
  79

orbital evolution   145

Painlevé's conjecture   21, 76, 78, 79
parametric plane   30
periodic orbit   122–126, 132, 134, 139
perturbation theory   70
perturbations   146
perturbed two body problem   63, 64
physical plane   30
planet crossing asteroid   183, 204
planet-crossing orbits   145
planetocentric velocity   145
Plummer   16
Potentially Hazardous Asteroids   8
proper elements   179, 203, 205, 207, 208
Pythagorean problem   21

quadratic form   189, 191

radial-inversion   25, 44, 47
regula falsi   201
regularization   49, 63
regularization by surgery   96
regularization theory   9
resonant orbits   145
restricted circular planar three body
  problem   49
restricted three-body problem   26
Runge-Kutta-Gauss method   201

scale separation   103
scaling transformation   83
secular resonances   204, 205
semianalytic   188, 204
Siegel's exponents   92
Siegel's series   90
singular perturbations   95
singularities   49, 72–75, 79
singularity   25, 32
Sundman's inequality   83
Sundman's theorem   84
symbolic dynamics   127, 129, 132
symplectic   201
synodic frame   32, 36
synthetic   204

tangent crossing   190
three-body problem   85
Torino impact scale   8
transition   122, 124, 134
triple collisions   17, 81
triple encounter   81
triple-collision manifold   96
triple-parabolic escape   94

Tunguska   5
two-body problem   26, 29

Vlasov equation   107
vortex dynamics   109

weak averaged solutions   188, 207
weak coupling limit   107
Wetherill   188–190, 192